ORDER AND DISORDER IN POLYMER REACTIVITY

ORDER AND DISORDER IN POLYMER REACTIVITY

G. E. ZAIKOV
AND
B. A. HOWELL
EDITORS

Nova Science Publishers, Inc.
New York

LIBRARY OF CONGRESS CATALOGING-IN-PUBLICATION DATA
Available upon request.

ISBN1-60021-145-3

Published by Nova Science Publishers, Inc. ✣ New York

CONTENTS

PREFACE

"He knew little,
but did much."
The opinion of Americans
on Ronald Reagan

These are well-liked and respectful words of American citizens about their president. Clearly, there are several alternatives of this expression:

1. Knew much and did much,
2. Knew much, but did little,
3. Knew little and did little.

Generally, the second or third expression describes people. Let it be so that criticism and self-criticism are shown.

Anton Chekhov – the famous Russian writer, once said that "a man is a fraction, where the denominator shows his personal estimation and the numerator shows his estimation by other people". Of course, if your personal estimation is low, the fraction will not be very small, even if a person has done little.

However, let us find out the readers' opinion about the authors of this collection and then deduce the fractions for different scientists.

Some part of this collection is devoted to an important topic: the order and disorder in polymers. This is a very important factor because it defines diffusion properties (molecule motion speed in polymeric matrix) and solubility of low molecular substances in polymers.

Recall that the diffusion coefficient in a glass of water equals $10^{-4} - 10^{-5}$ cm^2/s, in melted polymer – 10^{-10}, in a solid amorphous polymer – 10^{-15}, and in crystalline polymer – 10^{-21}-10^{-27} cm^2/s. Thus the molecule's motion and speed at transition from liquid low molecular compound to solid crystalline polymer may vary by 20 orders of magnitude or higher. Hence, a transition from kinetic area (when the reaction rate is defined by reactivities of the substances) to diffusion area (when the situation is completely defined by reagent delivery to the interaction site and the Frank-Rabinovich cage effect) may occur.

The rate of chemical processes in polymers strictly depends on the order degree in the polymer matrix. In its turn, this will affect the operation properties of polymeric materials and their storage life and reliable operation.

The authors would be grateful for positive comments on the materials of the current collection, which will be taken into consideration in our future work.

Prof. Bob A. Howell
Central Michigan University
Mount Pleasant, Michigan, USA

Prof. Gennady E. Zaikov
N. M. Emanuel Institute of Biochemical Physics
Moscow, Russia

INTRODUCTION: STABILIZATION OF VINYLIDENE CHLORIDE POLYMERS BY COMONOMER INCORPORATION

B. A. Howell
Center for Applications in Polymer Science
Central Michigan University
Mt. Pleasant, MI 48859

ABSTRACT

Vinylidene chloride polymers find important application in the barrier plastics packaging industry. These materials display low permeability rates for both oxygen (and other small molecules) and for food aroma and taste constituents. On the one hand, they function to prevent spoilage of packaged food items and, on the other, to prevent the loss of flavor agents that make these items palatable. While these materials have excellent barrier properties they may be processed only with difficulty owing to the propensity to undergo thermally-induced degradative dehydrochlorination. In fact, the homopolymer cannot be processed. Incorporating simple acrylate comonomers into the polymer structure lowers the melt temperature and improves processibility. However, it is insufficient to prevent significant degradation at process temperatures. Incorporation into vinylidene chloride polymers of a series of comonomers which result in the formation of polymer pendant groups with the potential 1.) to react with hydrogen chloride as it is formed (and thus prevent its interaction with the walls of process equipment to form Lewis acids, principally iron(III) chloride, which accelerate the dehydrochlorination reaction) and 2.) to expose phenolic units (which may scavenge chlorine atoms and other radical species) on reaction with hydrogen chloride has been examined as a means of stabilizing these materials.

The thermal degradation of poly(vinylidene chloride) and vinylidene chloride (VDC) copolymers usually occurs with the evolution of hydrogen chloride at elevated temperature[1]. For the homopolymer, the degradation occurs rapidly when heated to its melting point, making it difficult to formulate through extrusion processes. As a consequence, only the copolymers with vinyl chloride, alkyl acrylate and acrylonitrile, etc., are of commercial prominence [1]. These polymers have a lower melting point, and are more soluble in some solvent mixtures than the homopolymer so that processing is possible.

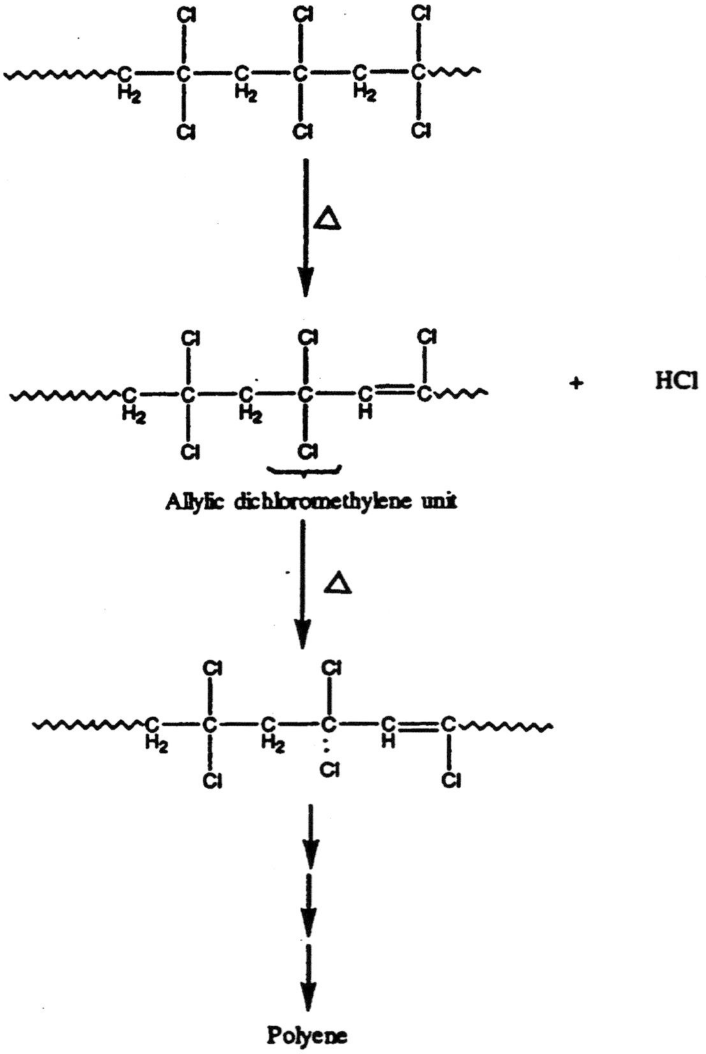

Scheme 1. Thermal Degradation of a Vinylidene Chloride Polymer

In the packaging industry, VDC copolymers have assumed a position of great importance because of the extremely low permeability for a wide variety of gases including oxygen and flavor and aroma constituents [2,3,4]. To retain these unique properties which are essential for high barrier to the mass transport of small molecules, the VDC content of these polymers is usually greater than 85%. There have been a number of studies focused on thermal decomposition processes of these polymers [5-10]. It has been suggested that a defect site, most probably internal double bonds introduced during the preparation or processing of the polymer is responsible for the initiation of the degradation [5-9]. The initiation, propagation and termination phases are typical for degradative dehydrochlorination along the polymer mainchain and the "unzipping" results in the formation of polyene sequences comprised of chloroacetylene units [5]. The "unzipping" can be stopped by comonomer units present such as acrylate, i.e., the length of the polyene sequences may be limited by the level of comonomer present, however, as noted above, the level of comonomer present cannot exceed 10-15% if the copolymer is to retain its barrier properties.

Therefore the copolymer must be stabilized in some way to control degradation during thermal processing [11]. As previously noted, internal double bonds are the defect sites responsible for the initiation of degradation. This is because the indroduction of a double bond (via dehydrochlorination) into a VDC sequence generates an allylic dichloromethylene which is very prone to carbon-chlorine bond homolysis to propagate the unzipping dehydroclorination reaction.

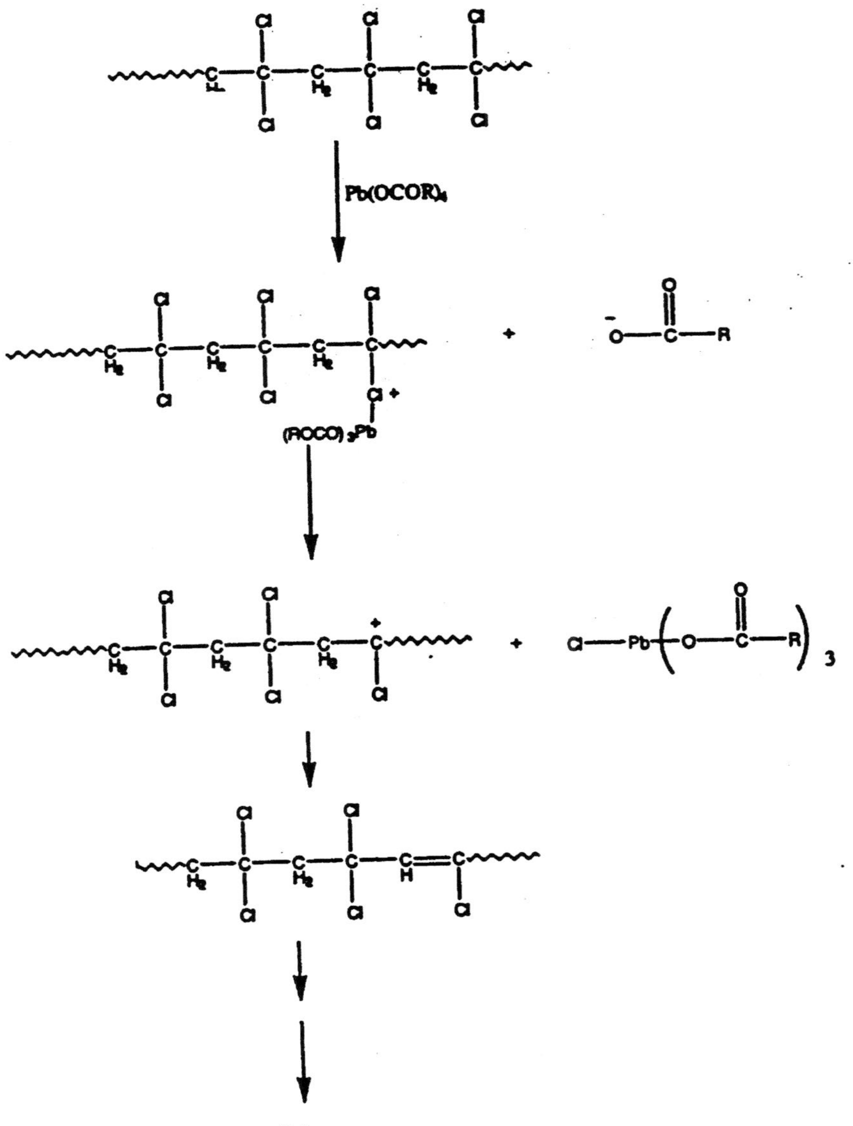

Scheme 2. Interaction of Metal Carboxylates with Vinylidene Chloride Polymers

One obvious approach to controlling the degradation sequence would be to treat the polymer with an agent which would replace allylic halogen with more thermally stable groups. This approach is successful for the stabilization of poly(vinyl chloride) (PVC) in which the polymer is treated with lead, tin, cadmium, or zinc carboxylates which convert allylic chloride groups to allylic esters wich degrade only above the temperature required for processing [12]. Treatment of vinylidene chloride polymers with these same compounds leads to rapid degradation of the polymers. This is due to the sensitivity of the dichloromethylene group to the presence of Lewis acids [13,14].

For this approach to be effective the metal cation must be capable of coordinating allylic halogen to promote displacement by carboxylate but not sufficiently acidic so as to promote carbon-chlorine bond heterolysis (as depicted in Scheme 2). Copper(II) carboxylate seems to posses the proper balance between Lewis acidity of the cation and nucleophilicity of the anion to behave as an effective stabilizer in these systems [15,16].

Similar stabilization can be effected with strongly nucleophilic but weakly basic additives. For example, amines, even highly hindered amines, are too basic to permit displacement of allylic halogen in the absence of E2-type dehydrohalogenation [17-19]. Trialkyl/aryl phosphates, on the other hand, have shown some utility as stabilizers in these polymers.

Other approaches to stabilization in these systems have involved the use of N-substitutedmaleimides to act as dienophiles to consume dienes as they are formed during propagation of the dehydrochlorination reaction or radical scavengers (hindered phenols) to trap chlorine atoms formed from the homolysis of allylic carbon-chlorine bonds [20,21,22]. A passive base such as magnesium oxide or tetrasodium pyrophosphate is often a component of a suitable stabilization package as well [23]. The purpose of these agents is to scavenge hydrogen chloride as it is formed and prevent its interaction with the walls of processing equipment to generate Lewis acids, particularly iron(III) chloride, which promote the dehydrochlorination reaction.

A potentially more useful approach which has received much recent attention has been the incorporation into the polymer of a comonomer which contains a substitutuent group suitable for reaction with hydrogen chloride to expose a phenolic unit potentially capable of scavenging chlorine atoms [24-27].

In the first instance, 4-acetoxystryrene was used as a comonomer to produce a copolymer containing acetoxy groups [24]. Reaction of the acetoxy group with evolved hydrogen chloride could consume a mole of hydrogen chloride to liberate acetic acid (a weak acid) and a phenol which might scavenge chlorine atoms. Unfortunately, under the conditions of thermal degradation of the polymer the interaction of anhydrous hydrogen chloride with the ester was not sufficient to effect cleavage. Therefore, the presence of acetoxystyrene units in the polymer did not impart stability. Indeed, the introduction of a benzylic group into the polymer mainchain provided a site for thermally-induced chain scission and led to a copolymer that was considerably less stable than the corresponding acrylate copolymer. Similarly, the generation of a copolymer containing pendant ether functionality (catechol cyclopentanone ketal) was not a useful approach for the synthesis of a stable vinylidene chloride copolymer [25]. Again the presence of anhydrous hydrogen chloride was not sufficient to promote cleavage of the ketal.

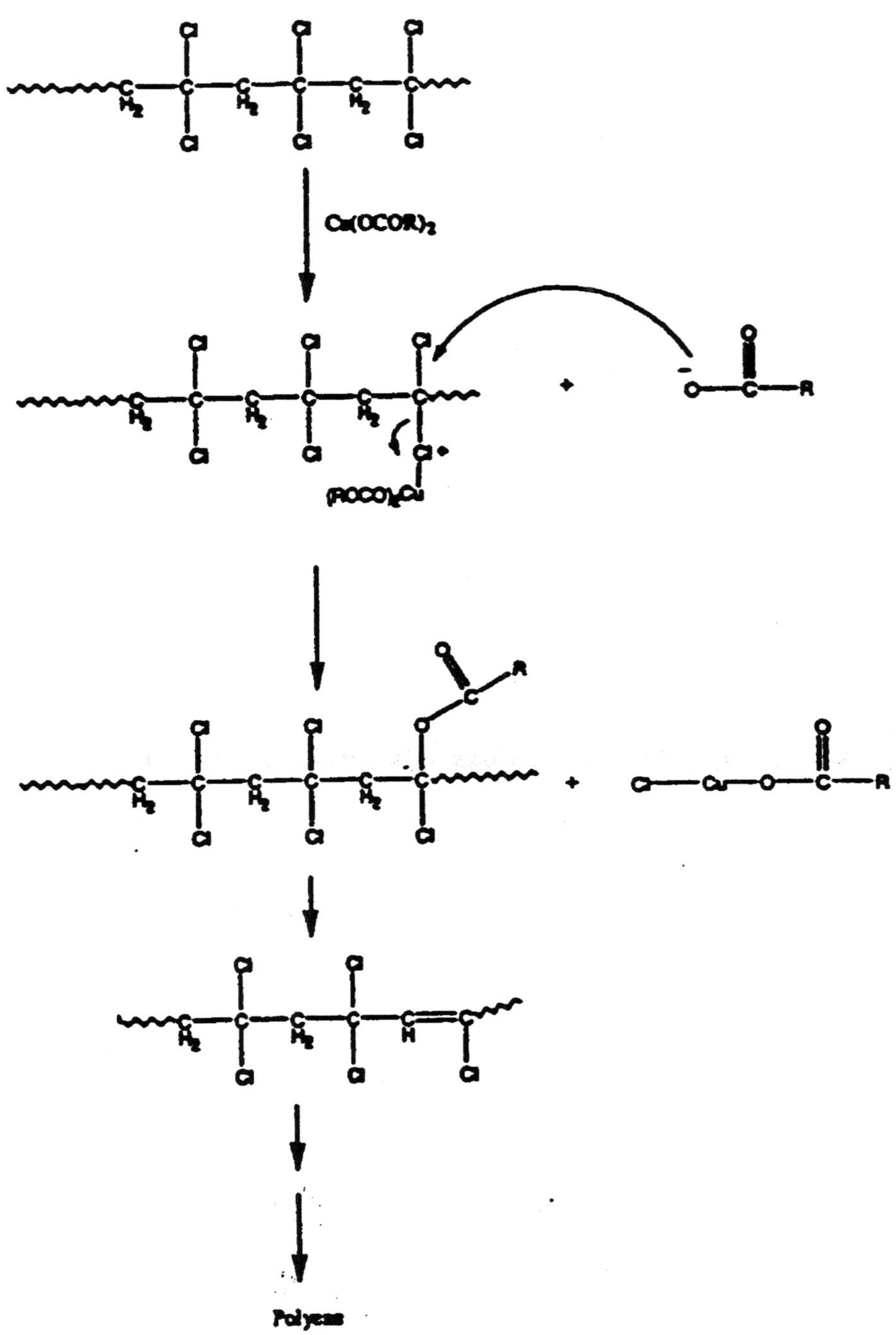

Scheme 3. Interaction of Copper(II) Carboxylate with a Vinylidene Chloride Polymer

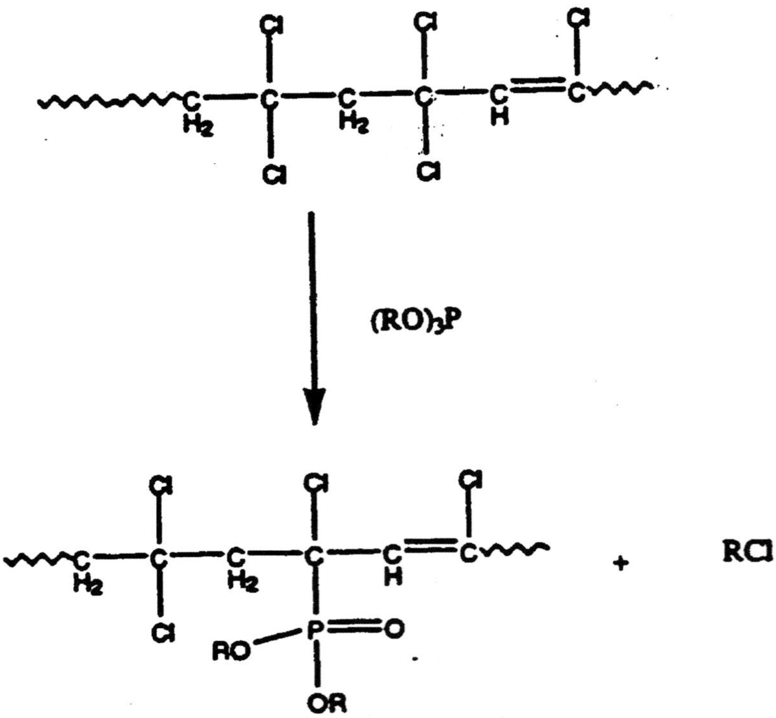

Scheme 4. Stabilization of Vinylidene Chloride Polymers by Trialkyl/aryl Phosphites

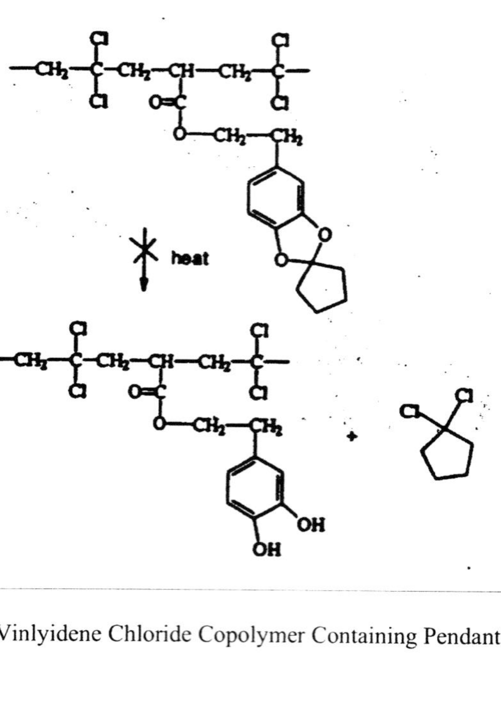

Scheme 5. Degradation of a Vinlyidene Chloride Copolymer Containing Pendant Catechol Cyclopentanone Ketal Functionality.

In contrast, the carbonate moiety is susceptible to cleavage under conditions of copolymer degradation. A copolymer containing a benzyl acrylate bearing a *t*-butoxycarbonyloxy group in 4-position of the phenyl nucleus undergoes thermal degradation to evolve several volatile products in addition to hydrogen chloride [26,27]. Infrared analysis of the decomposing polymer suggested that the acrylate units were being converted to acrylic acid units, i.e., that the polymer was being converted to a copolymer of vinylidene chloride and acrylic acid. The decomposition may be depicted as shown below.

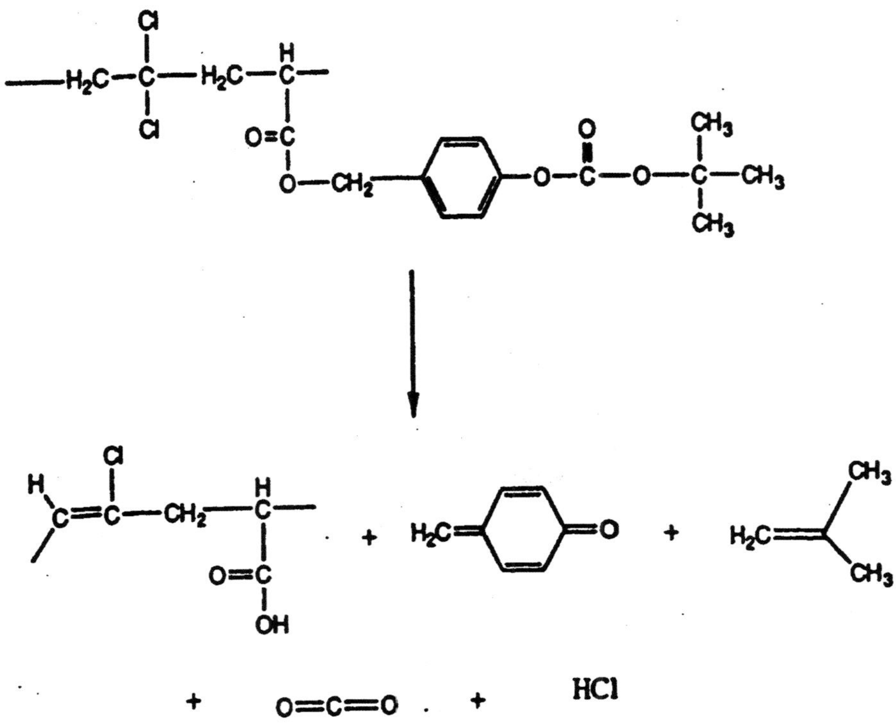

Scheme 6. Thermal Degradation of a Vinylidene Chloride/[4-(t-Butoxycarbonyloxy)phenyl]methyl Acrylate.

CONCLUSIONS

Vinylidene chloride copolymers are important materials for use in barrier plastic packaging applications Because of the propensity to undergo thermally-induced degradative dehydrochlorination during processing these polymers must be stabilized. The homopolymer, poly(vinylidene chloride), undergoes catastrophic dehydrochlorination at its melt temperature (200° C) and is not a commercial material. Incorporation of an alkyl acrylate comonomer (4-10%) into the polymer lowers the melt temperature and limits the vinylidene chloride sequence length. The resulting polymers can be processed and are extremely good barrier resins for food packaging applications. The incorporation of acrylate units containing pendent groups capable of absorbing hydrogen chloride to expose a moiety that can scavenge chlorine atoms and other radical species has the potential to provide resins of unusual stability.

REFERENCES

[1] R.A. Wessling, D.S. Gibbs, P.T. Delassus, D.E. Obi, and B.A. Howell, *Kirk-Othmer Encyclopedia of Chemical Technology*, Vol. 24, 2nd Ed., John Wiley and Sons, Inc., New York, NY, 1997, pp. 883-923.

[2] P.T. Delassus, W.E. Brown, and B.A. Howell in A.L. Brody ad K.S. Marsh, Ed., *Enclyclopedia of Packaging Technology*, 2nd Ed., John Wiley and Sons, Inc., New York, NY, 1997, pp. 958-961.

[3] G. Strandburg, P.T. Delassus, and B.A. Howell, in S.J. Risch and J.H. Hotckiss, Ed., *Food and Packaging Interaction II*, American Chemical Society (Symposium Series No. 473), Washington D.C., 1991, Ch.12.

[4] P.T. DeLassus, G.Strandburg and B.A. Howell, *Tappi J.*, 71, 177 (1998).

[5] B.A. Howell, *J. Polym. Sci., Polym. Chem. Ed.*, 25, 1681 (1987).

[6] B.A. Howell, P.T. Delassus and C. Gerig, *Polym. Prepr.*, 28(1), 278 (1987).

[7] B.A. Howell and P.T. Delassus, *J. Polym. Sci., Polym. Chem. Ed.*, 25, 1697 (1987).

[8] B.A. Howell and P.B. Smith, *J. Polym. Sci., Polym. Chem. Ed.*,26, 1287 (1988).

[9] B.A. Howell, *Thermochim. Acta,*134, 207 (1988).

[10] S. Collins, K. Yoda, N. Anazawa, and C. Birkinshaw, *Polym. Degrad. Stability*, 66(1), 87, 93 (1999).

[11] B.A. Howell, B.S. Warner, C.V. Rajaram, S.I. Ahmed and Z Ahmed, *Polym. Adv. Tech.,* 5, 485 (1994).

[12] T. Hjertberg, E. Martinddon and E. Sorvik, *Macromolecules,* 21, 603 (1998).

[13] B.A. Howell and J. R. Keeley, *Thermochim. Acta,* 272, 131 (1996).

[14] B.A. Howell and A.Q. Campbell, *Thermochim. Acta*, 340, 231 (1996).

[15] B.A. Howell and C.V. Rajaram, *J. Thermal Anal.*, 40, 575 (1993).

[16] B.A. Howell and C.V. Rajaram, *J. Vinyl Tech.,* 15, 202 (1993).

[17] B.A. Howell and H. Liu, *J. Vinyl Tech.*, 13, 187 (1991).

[18] B.A. Howell and H.Liu, *Thermochim. Acta*, 212, 1 (1992).

[19] B.A. Howell and F.M. Uhl, *Thermochim. Acta,* 357, 127 (2000).

[20] B.A. Howell and J. Zhang, *Polym. Prepr.*, 42(2), 624 (2001).

[21] B.A. Howell, M.F. Debney, and C.V. Rajaram, *Thermochim. Acta,212, 215 (1992).*

[22] B.A. Howell, Z. Ahmed and S.I. Ahmed, *Thermochim. Acta*, 357, 103 (2000).

[23] B.A. Howell, F.M. Uhl and D. Townsend, *Thermochim. Acta*, 357, 127 (2000).

[24] B.A. Howell and R.C. Mason, unpublished results.

[25] B.A. Howell, B.B.S Sastry, S.I. Ahmed and P.B. Smith, *Thermochim. Acta*,272, 139 (1996).

[26] B.A. Howell, D.A. Spears and P.B. Smith, *Thermal Degradation of Vinylidene Chloride/[(4-t-Butoxycarbonyloxy)phenyl]methyl Acrylate Copolymers,* 24th North American Thermal Analysis Society Meeting, San Franscisco, CA, September, 1995, No. 60.

[27] B.A. Howell and B. Pan, *Polym. Mater. Sci. Eng.,* 76, 401 (1997).

In: Order and Disorder in Polymer Reactivity
Editors: G. E. Zaikov and B. A. Howell, pp. 1-12
ISBN 1-60021-145-3

Chapter 1

Degradation and Stabilization of Liquid Crystalline Polymers

E. V. Kalugina[1], A. B. Blumenfeld[2], Ja. G. Yrman[2], A. L. Narkon[2], K. Z. Gumargalieva[3] and G. E. Zaikov[4]

[1] Research and Production Company "Polyplastic", 14a, General Dorochov street, Moscow, 119530
[2] G. Petrov's Institute for Plastics, 35, Perovskiy proezd, Moscow, 11124
[3] Institute of Chemical Physics, Russian Academy of Sciences, 4 Kosygin Street, Moscow 117334 Russia
[4] Institute of Biochemical Physics, Russian Academy of Sciences, 4 Kosygin Street, Moscow 117334 Russia

Abstract

Some liquid crystalline copolyesters (LCP) derived from dihydroxydiphenylacetate, acetobenzoic acid, isophthtalic and terephthtalic acids with various proportions of the latter two comonomers were investigated. The LCP thermal stability and the degradation behavior under processing temperatures were studied. The kinetics of loss in weight, the oxygen uptake as well as the oxidation product evolution were studied. On the basis of the data obtained a mechanism for the LCP high-temperature thermal oxidation, relative to such for other polyheteroarylenes was proposed.

Key words: Liquid crystalline copolyesters, hydroxybenzoic acid, phenolic antioxidant, inhibits thermal oxidation, peroxy radicals, phosphate.

Materials

Liquid Crystalline Copolyesters (LCP)

Laboratory products synthesized from dihydroxydiphenylacetate, acetobenzoic acid, isophthtalic acid (IPA) and terephthtalic acid (TPA). The TPA/IPA ratio (mol %) used for the copolymer preparation were 100/0, 75/25, 50/50, 25/75 or 0/100 (LCP-1, LCP-2, LCP-3, LCP-4 and LCP-5 correspondingly).

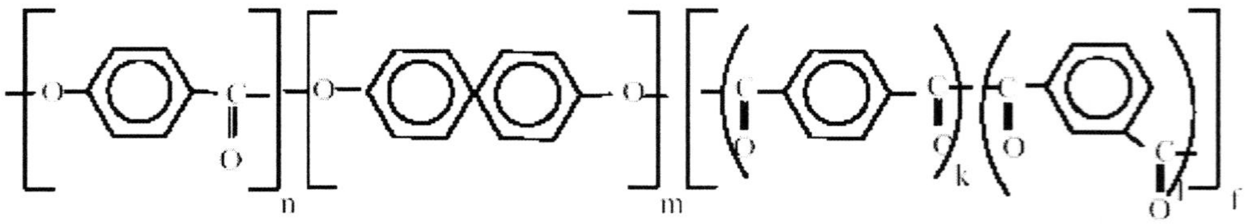

Thermal Stabilizers

Commercial products Irganox 1010, Irgafos 168, Irgafos 126, Irgafos P-EPQ (Ciba-Geigy).

Methods

^{1}H, ^{13}C and NMR- Spectroscopy

NMR-spectra were recorded on a NMR spectrometer Gemini-300 (Varian). The polymers and the degradation products were dissolved in pentafluorophenol or in chloroform and analyzed. Assignments for chemical shifts were performed using MNR-spectra catalogues [1-4].

IR-Spectroscopy [5-7]

The virgin LCP and the LCP after heat treatment as well as the degradation products were analyzed. IR-spectra were recorded on a Fourier-IR-spectrometer (Perkin-Elmer, model 1700) and a spectrometer Specord M80 (Karl Zeiss).

Mass-Spectrometry

Mass-spectrometry in the mode of "direct" input of volatile products into the ionization camera of a spectrometer M-80A (Hitachi) operated at a temperature program up to 400°C was used. The peak identification was fulfilled using mass spectra data banks.

Gas Chromatography

Volatile degradation products were analyzed using a gas chromatograph "Zvet-168"(Russia). Detector – catarometer, columns with molecular sieves CaX and Poropac Q. A special input device for the samples sealed in ampoules was used.

Thermal Analysis

TGA and DTA were performed in argon or in air flow under dynamic (5 °/min) heating. A simultaneous thermal analyzer STA-781 (Stanton Redcroft) was used.

ICP Emission Spectroscopy

A sequential inductively coupled plasma (ICP) emission spectrometer "Atomscan-25" (Thermo Jarrel Ash) was used to measure the metal content in the polymers. Test solutions were prepared by mineralization of test portions (0,5 - 1 g) of the polymers under interest. A mixture of nitric acid and perchloric acid was used for mineralization. Test solutions were aspirated into the argon plasma and analyzed. Calibration for the metals to be determined was performed using certified standard solutions.

X-Ray Structure Analysis (RSA)

An X-ray diffractometer Dron-3 (Russia) operated at the temperature scanning between 20°C and 400°C was used. Powdery or injection-molded samples were investigated.

Compounding

Powdery LCP were mixed with additives and processed using a twin screw vacuum extruder Collin, L/D=25.

Results and Discussion

Liquid crystalline polymers belong to a class of thermally stable engineering thermoplastics. They exhibit excellent physical and mechanical properties up to 350°C [8], low linear thermal expansion, high dimension stability, chemical resistance and low combustibility. They differ from common thermoplastics by ability to self-reinforcing [9-12]. Progress in computer technologies provoked a special interest in these polymers. A number of works on the LCP structure and properties were published during the last 5 – 10 years. Nevertheless, there is little information about the LCP degradation behavior.

STRUCTURE AND THERMAL STABILITY OF LCP

Thermal stability of LCP as a function of the TPA/IPA proportion is evaluated with the help of dynamic TGA/DTA under inert and oxidizing atmosphere. Melting intervals and the points of the degradation beginning are presented in Table 1. The TGA curves obtained at heating in air are presented in Fig.1.

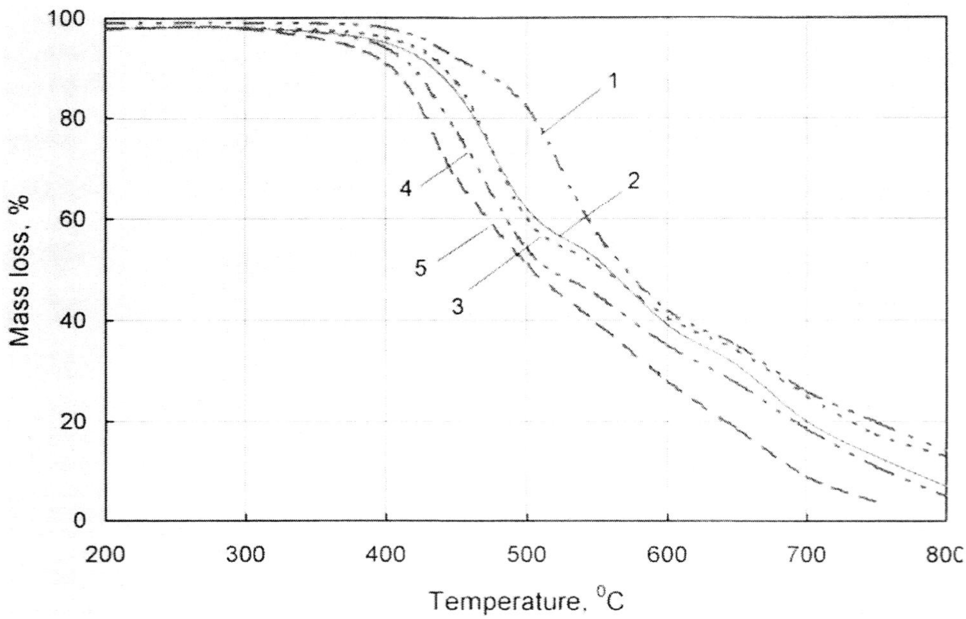

Figure 1. TGA-date in air at 5 o/min for LCP-1 (1), LCP-2 (2), LCP-3 (3), LCP-4 (4), LCP-5 (5).

Table 1. TGA/DTA in argon flow (heating rate 10°/min)

Polymer	TPA/IPA ratio, mol %	Melting interval, °C	Degradation start temperature, °C
LCP-1	100/0	410-415	490
LCP-2	75/25	360-365	480
LCP-3	50/50	340-350	472
LCP-4	25/75	300-315	455
LCP-5	0/100	315-325	450

The LCP decomposition in the absence of oxygen proceeds in one step, a coke formed at 700°C is about 40% (w/w). There are two endotherms on the DTA curves: a weak peak corresponding to the polymer melting (ΔH=1 - 2.5 KJ/mole) and an intensive one revealing the sample decomposition. Having investigated a copolymer of 4-hydroxibenzoic acid and 2,6-hydroxynaphthoic acid by DSC, the authors [13] supposed that $\Delta H \sim 1$ KJ/mole revealed the change in ordering during the crystalline–mesomorphic phase transition rather that true melting. Unfortunately, neither these workers neither we have not been able to catch the transition connected with the LCP isotropization. This is probably because of superimposition of heat effects that accompany the degradation process.

The LCP investigated indicate worse thermal stability in air than in argon. The dynamic TGA curves (Fig. 1) showed weight losses in air 25-30° below than that of in inert atmosphere. Decomposition proceeds in two steps. Up to 40 % (w/w) of the sample is lost during the first step (350°C -550°C). Practically complete decomposition of the polymers is observed during the second, slower stage (550-800°). The coke residue is 3-10% (w/w) at 750-800°C. According to DTA, the LCP decomposition in air is accompanied by heat evolution. The endothermic effect which is characteristic for the LCP thermal degradation is absent under the oxidizing conditions. This is probably due to this peak is overlapped by intensive exotherms. The LCP thermal stability diminishes in the row: LCP-1 > LCP-2 ≥ LCP-3 > LCP-4 > LCP-5.

The increase in the IPA fragments content shifts the "melting" interval to lower temperatures and reduces the polymer thermal stability.

The investigation of the phase transitions, occurring in LCP at 20°C - 400°C, by RSA shows relative alternations. Annealing at 300°C slightly intensifies the main crystalline reflection. The diffractograms for the powdery and injection-molded LCP-2 are given in Fig.2 as an example. In the temperature interval investigated we found out no changes (appearance or disappearance of reflections) in the polymers. This confirms the DTA results as well as the proposal on the neighborhood of the degradation and the mesomorphic phase formation temperatures [13]. The behaviors of the materials under processing temperatures reveal their ability to transfer to so-called liquid crystalline state. All the polymers investigated display a jump drop in viscosity at the softening points (300°C - 400°C, depending on the structure), and very strong fibers are formed from the melt. This has been interpreted [8] by cooperative orientation of the main axes of macromolecules along the flow direction (viscosity anisotropy) that is characteristic only for LCP.

As known, the polymer thermal stability depends on many factors including the structure, the molecular weight characteristics, the concentration of defects and labile end groups (hydroxyl-groups in this case) in macromolecules as well as on low-molecular weight admixtures (for example, monomers, catalysts, metal ions from raw materials and equipment).

We measured the concentration of inorganic admixtures in the LCP and appropriate monomers by ICP emission spectrometry. Taking LCP-2 as an example, we tried then to compare the metal content with the polymer thermal stability. To clarify the effect of metal impurities we have increased their concentrations artificially by introduction of appropriate inorganic salts. The data obtained for the effect of Cu, Fe, Ni, Ca and Al on the decomposition onset temperature T_d according to TGA are listed in Table 2.

As can be seen, the metal admixtures affect the polymer thermo-oxidative stability in different ways. There is no visible effect of aluminum, alkali or alkali-earth metals (Na, K, Ca); Fe lows, but Cu and Ni in the concentration up to 10^{-3}-10^{-4} % (w/w) improve the LCP thermal stability.

Organic admixtures in LCP were searched by mass-spectrometry. Phenol and dihydroxydiphenyloxide (the product of the dihydroxydiphenylacetate hydrolysis, m/e 94, 186) as well as the following fragments with higher molecular weight were identified.

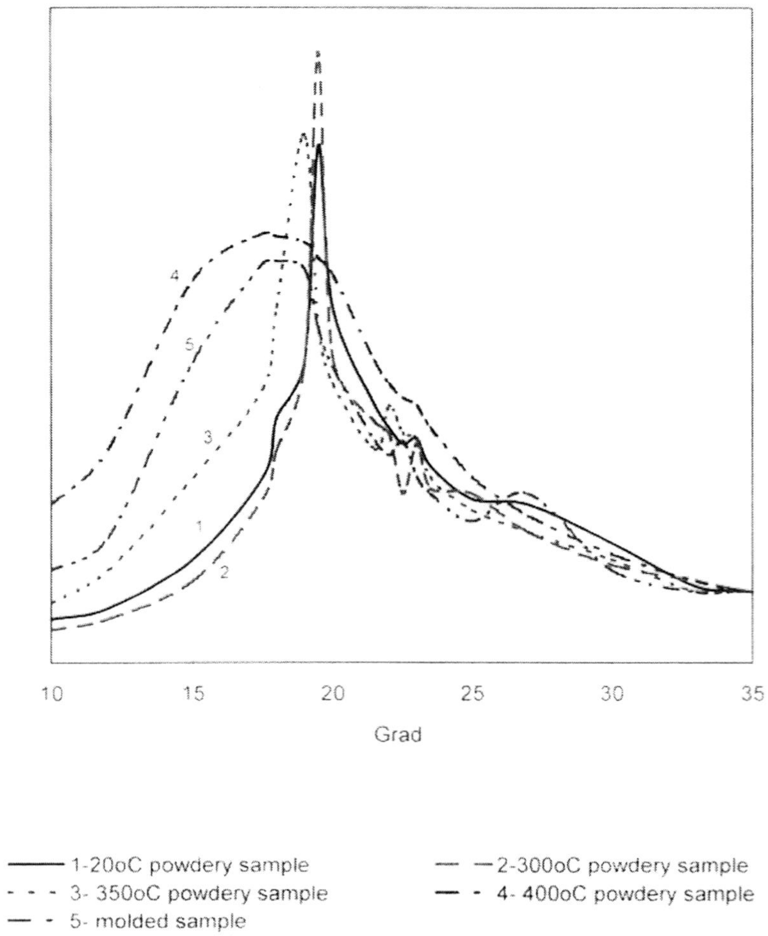

Figure 2. The diffractograms for the powdery LCP-4 at heating.

Table 2. Effect of the metal admixtures on the LCP-2 thermo-oxidative stability

Metal	Content, % (w/w)	T_d, °C (TGA in air, heating rate 5°/min)
Fe	1.3 10^{-3}	320
	1.3 10^{-2}	300
Al	1.4 10^{-3}	320
	1.4 10^{-2}	320
Ca	4 10^{-3}	320
	4 10^{-2}	320
Ni	1.0 10^{-4}	320
	2.0 10^{-2}	325
Cu	1.3 10^{-5}	320
	2.0 10^{-3}	330
Ni	1.0·10^{-5}	320
	2.0·10^{-3}	330

The total amount of organic moieties in samples is not more than (1.0-2.0) 10^{-2} % (w/w) that practically does not affect the polymer thermal stability.

By the oxygen uptake kinetics, the LCP thermal stability at 350°C (processing temperature) decreases when the IPA content grows (Fig.3.1). This is confirmed by the TGA data. The fast O_2 uptake corresponding to the first fast stage (2-3 h.) is about 1 mole per monomeric unit. During the second, slow stage only 0.2 mole/monomeric unit is burned up over 7-8 h. Oxygen dioxide, the main volatile product of the LCP thermal oxidation, evolves in a similar manner. (Fig.3.2). The heavy high-boiling degradation products, which are identified by mass-spectrometry and NMR, are the following:

Heavy Degradation Products

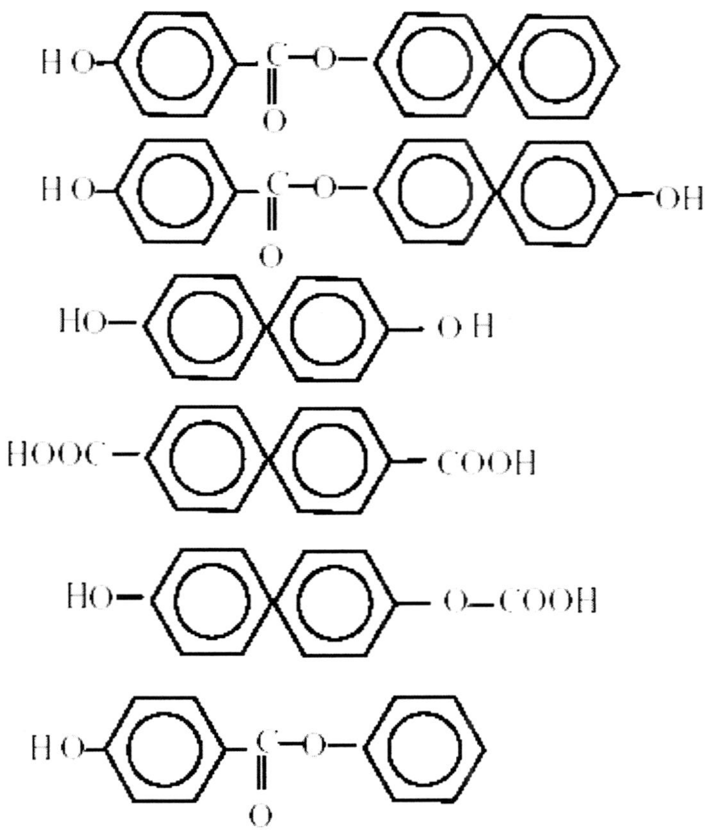

m/e=306

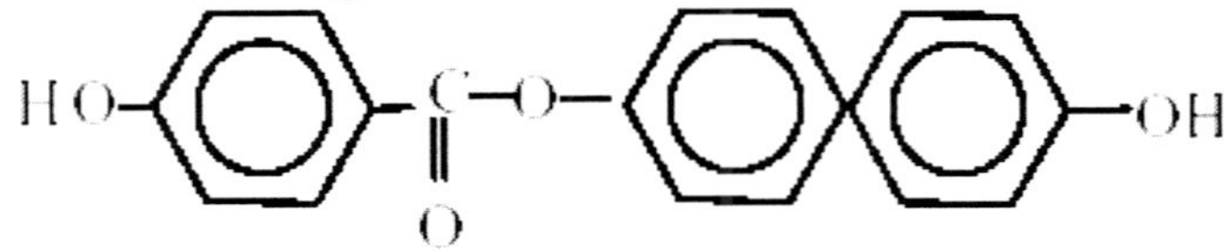

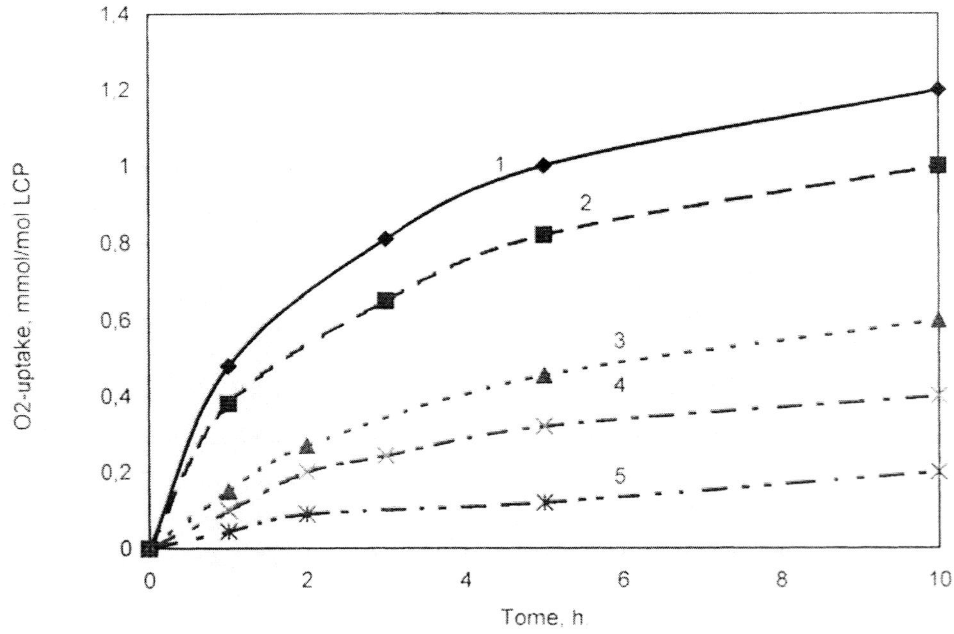

Figure 3.1. Kinetic curve for O2-uptake at 350 0C in air of LCP-5(1), LCP-4 (2), LCP-3(3), LCP-2(4), LCP-1 (5).

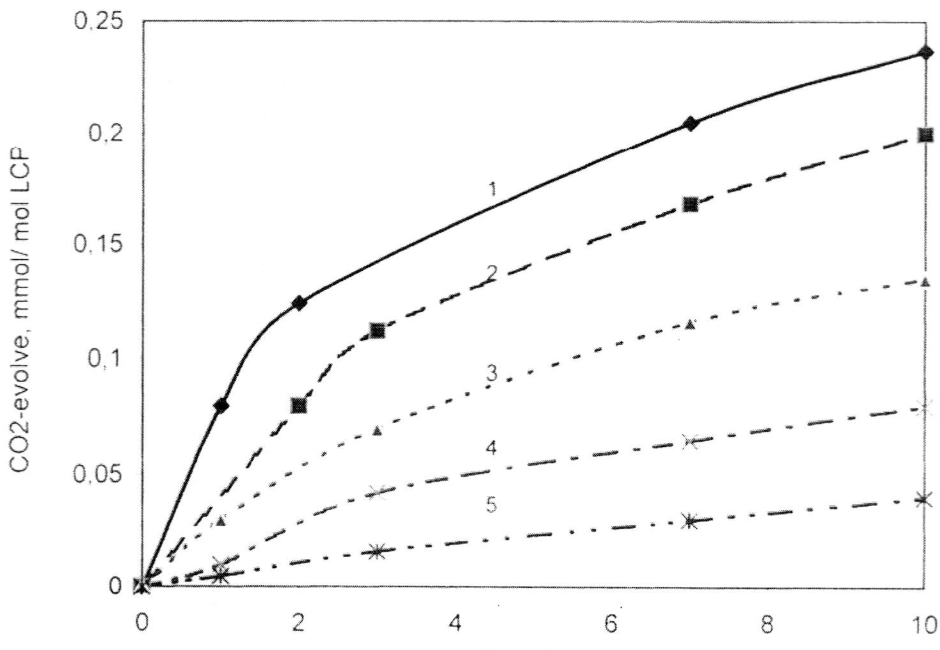

Figure 3.2. Kinetic curve for CO2-evolve at thermal oxidation (350 0C, air) LCP-5 (1), LCP-4 (2), LCP-3 (3), LCP-2 (4), LCP-1 (5).

These nonvolatile products deposited on the cold walls of ampoules are a mixture of dihydroxydiphenyldiacetate (starting monomer) with some substances formed due to total or partial hydrolysis of the monomers (for example, dihydroxydiphenyl) as well as due the dihydroxydiphenyl and hydroxybenzoic acid interaction. Notice that the degradation product composition is the same for the LCP investigated, but their relationship differs.

Taking LCP-2 as an example, we examined the behavior of the polymer under the processing conditions in more detail. The thermal oxidation rate at 300 и 320°C (near the melting point) is noticeably less than at 350°C (see Fig. 4.1, 4.2)., when, by the RSA data, the polymer is completely transformed into the isotropic melt. Along with carbon dioxide, the main volatile degradation product Hydrogen and water were detected at the LCP-2 oxidation. Hydrogen evolves at the early stages of thermal oxidation (0.5-1.0 h), the water content is measurable for the exposure more than 4 hours. Chemical analysis shows that the hydrogen content in LCP-2 drops in the coarse of thermal oxidation whereas the carbon content in the residue grows, that is, a graphite-like structure is formed. This process is rather intensive at 350°C. The IR spectra show that the intensity of the peaks corresponding to the aromatic ester fragments ($\nu_{C=O}$= 1740 cm^{-1}, ν_{C-O}=1270, 1160 cm^{-1}, $\nu_{C=C}$=1600, 1500 cm^{-1}, $\delta_{C=C}$=720 cm^{-1}) decreases first of all. After ten hours at 350°C only the bands assigned to ether bonds (ν_{C-O-C}=1080 cm^{-1}) and to an aromatic-type structure persist; the spectra background significantly decreases, that may be connected with crosslinking. Heating at 350°C is accompanied by evolution of oligomers. These heavy products are deposited on the ampoule walls (at a temperature near 150°C). The structure of the oligomeric degradation products having corresponding end groups is identified by ^{13}C NMR:

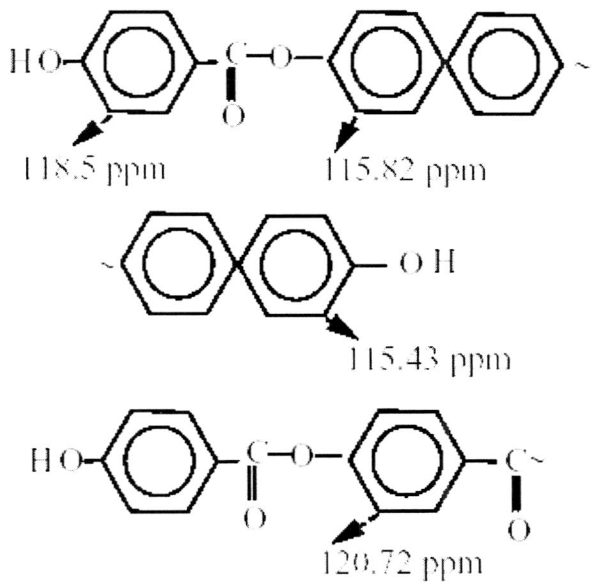

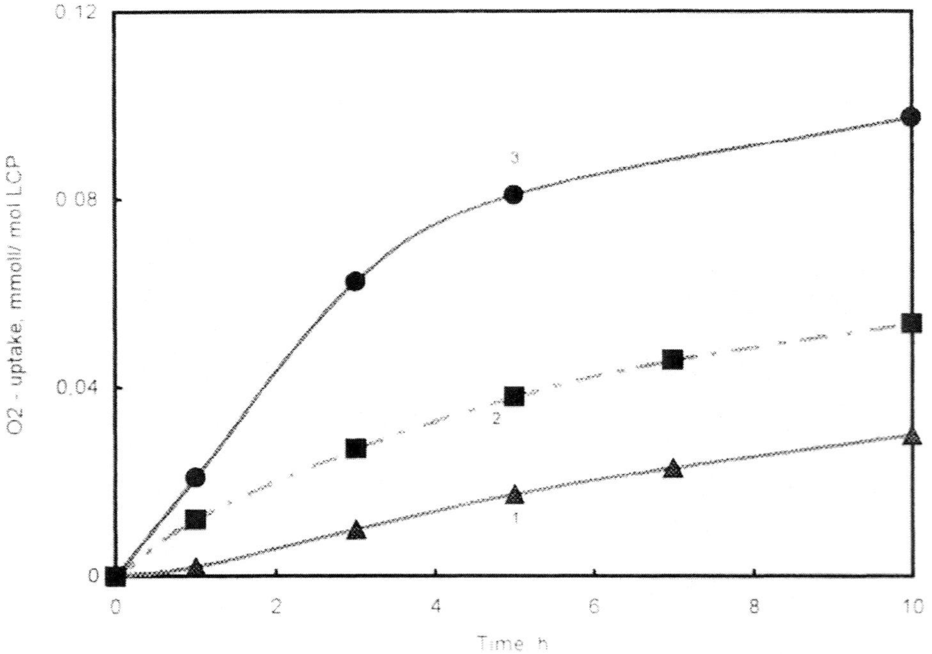

Figure 4.1. Kinetic curve for O2-uptake of LCP-4 at 300 (1), 300 (2), 350 (3) oC in air.

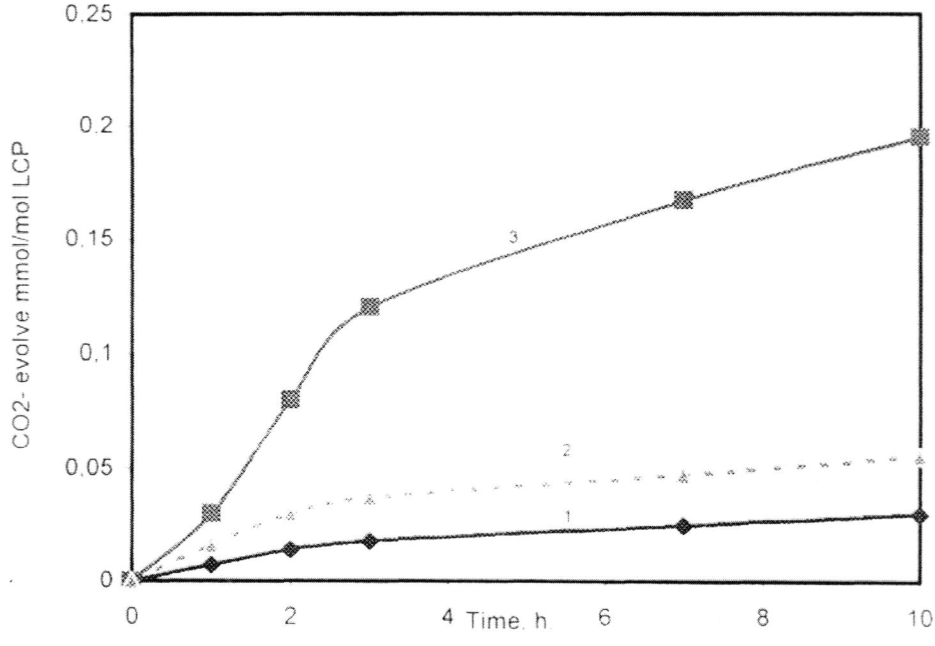

Figure 4.2. Kinetic curve for CO2-evolve at thermal oxidation (350 0C, air) LCP-4 at 300 (1), 320 (2), 350 (3) 0C.

The ratio of the peak areas allows estimating the oligomeric product concentration in the mixture as follows:

product A (118.5 и 115.82 ppm) - 15 mol %,
product B (115.43 ppm) – 1 mol %
product C (120.72 ppm) - 10 - 2 mol %

The presence of the oligomer with adjacent fragments of hydroxybenzoic acid (see product C) in the heavy degradation products makes it possible to assume that during the LCP synthesis homopolycondensation of p-acetoxybenzoic acid proceeds along with the comonomer co-polycondensation. The amount p-acetoxybenzoic acid found in the degradation products is 3-4 times higher than the amount of acetobenzoic end groups in the oligomers evolved. Apparently, the free p-hydroxybenzoic acid is formed as a result of the rapture of labile bonds in p- oxybenzoic sequences.

The comparison of the results obtained in this study for the LCP degradation under processing temperatures with the peculiarities of some thermally stable polyheteroarylenes degradation [14] brings to light some common features: carbonization of the structure, H_2 evolution, improvement in thermo-oxidative stability with transition metal compounds. That is why we took into account the stabilization of polysulfones, aryl-aliphatic polyimides, polyamides *etc*. The approach to such stabilization is based on the following proposals on the mechanisms of the above said polymer degradation:

1. Classical free radical chain mechanism of thermal oxidation;
2. O_2 -macromolecule complex formation,
3. Transition of macromolecules to an electron exited state.

Introduction of additives is a well-known method for the mechanism of the chemical reaction investigation. The experience has shown effectiveness of triple stabilizing formulations (Cu-containing compound + phenolic antioxidant + phosphite) for aryl-aliphatic polyamides, polyphthtalamides and other thermally stable polymers. The conception of such formulation is based on the mechanisms of the stabilizing action of the components:

- phenolic antioxidant inhibits thermal oxidation by reaction with peroxy radicals ROO•;
- phosphite, a secondary antioxidant, destroys hydroperoxides;
- Cu^{2+}- containing substance affects in different ways, for example inhibits peroxyradicals via a complex or reacts with defects of macrochains.

The effect of a stabilizing formulation on the LCP-2 thermal oxidative stability demonstrates Table 3.

Table 3. Thermal oxidation of LCP-2 in air (350°C, 30 min)

Stabilizing formulation	O_2 uptake, mole/monomeric unit	CO_2 evolution, mole/monomeric unit
Unstabilized	0.22	0.033
$CuSO_4$ (0.005%) + Irgafos 126 (0.3%) + Irganox 1010 (0.1%)	0.08	0.019

The results show that stabilization by additives makes possible a significant (more than twofold) retardation of the thermal oxidation reactions proceeding in LCP.

Acknowledgments

The authors are deeply thankful to the specialists of the RSA laboratory of the Research Institute for Plastics for the help in the work and the result discussion.

References

[1] Ja. Slonim, Ja.G.Urman *NMR-spectrometry for heterochain polymers*, Moscow, Chemical, 1982.

[2] L.F. Johnson, Jankowski. *Carbon-13 NMR spectra*, Wiley-Interscience Publ., N.Y.-London-Sidney-Toronto, 1972.

[3] E. Breitmaier, G. Haas, W. Voelter. *Atlas of Carbon-13 NMR data*, Heyden, London-Philadelphia-Rheine, 1979.

[4] V. Formacek, L. Desnoyer, H.P. Kellerhals, T. Keller, J.T. Clerc. ^{13}C Data ank, *Bruker Physik,* Karlsruhe,1976.

[5] Koji Nakanishi *Infrared absorption spectroscopy*, Holden-Day, Inc., San Francisco and Nankodo Company Limited, Tokyo, 1962..

[6] Hummel/Scholl. *Atlas of polymer and plastics analysis.*

[7] *Ultrarotspektroskopische untersuhungen an polymeren.* Von Johannes Dechant unter mitarbeit von Rudi Danz Wolfgang Kimmer, Rudolf Schmolke, Akademie-Verlag-Berlin, 1972.

[8] N.A.Plate *Liquid Crystalline polymers*. Moscow, Chemie, 1988.

[9] .*Kunstoffe* – 1990-20, N10, s.1159

[10] *Kunstoffe* – 1991-21, N9, s.214

[11] ANTEC'90: *Plast. Environ Jesterday. Today and Tomorrov. Soc. Plast. Eng. 48th Anny. Techn. Conf. And Exib.*, Dallas, Tex., May.

[12] *J. Appl. Polym. Sci.*- 1990-41, N 11-12, p.2723

[13] *J. Polym. Sci. Polym. Plus. Ed.*- 1985, v.23, N 4, p.521

[14] E.V.Kalugina, T.N.Novotortceva Liguid crystalline copolymers as modifiers for some thermoplastics. *Russian Plastics.*

In: Order and Disorder in Polymer Reactivity
Editors: G. E. Zaikov and B. A. Howell, pp. 13-18
ISBN 1-60021-145-3

Chapter 2

Influence Hexsaazocyclanes on a Microstructure of Polyethylene Terephthalate

O. V. Burykina and F. F. Nijazi
Kursk State Technical University, 94, 50 years of October Kursk, 305040, Russia

Abstract

Research of influence of introduction of hexsaazocyclanes on a microstructure of polyethylene terephthalate is carried out. The increase in ability modified PETP to crystallization, in comparison with not modified PETP, as hexsaazocyclanes play a role of the original centers of crystallization is revealed. Research plasticization effect of hexsaazocyclanes on polyethylene terephthalate is carried out. It is revealed, hexsaazocyclanes introduction influences a maximum of a tangent of dielectric losses polyethylene terephthalate, displacing it aside smaller temperatures.

Introduction

One of the most effective ways of new polymeric materials, obtaining with the properties, given before is modification of known polymers by target additives.

Changing chemical structure and quantities of the additive – the modifier, it is possible to improve in a proper way, those operational properties which are required for the given consumer.

Experimental Part

Hexsaazocyclanes (HZ), were taken as additives, which have the advanced circuit of interface, high photo stability and thermo stability.

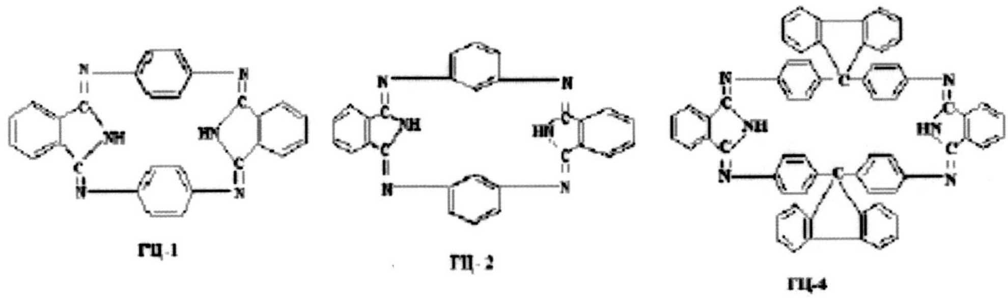

Figure 1. Structural formulas used hexsaazocyclanes

Availability chromatic groups, allows to use them in addition as dyes of polymers [1]. Proceeding from this, it was represented to us expedient, to compare thermal characteristics of additives and initial polymer, and also to define compatibility of these connections with fusion PETP.

Characteristics of the used additives of modifiers are submitted in table 1.

Table1. Characteristics hexsaazocyclanes

Designation	Average molecular weight	$\lambda_{макс.}$, nanometer (solvent acetone)	Painting of dye	T fusion, ^{0}C	T the beginnings of destruction ^{0}C
ГЦ-1	600	360	yellow	305	330
ГЦ-2	580	380	citric	327	345
ГЦ-4	454	400	beige	345	330

Temperatures of fusion and the beginning of decomposition hexsaazocyclanes are in limits T_{fusion}. = 305-345^{0}C and $T_{destruction}$. = 330-345^{0}C, that speaks about a potential opportunity of introduction of the given connections in fusion PETP as their thermo stability surpasses temperature of reception and formation PETP.

Polyethylene terephthalate modification hexsaazocyclanes by additives was carried out and fibers were formed. Additives introduction did not influence on the process of formation, and, consequently on the polymer fiber properties.

Discussion of Results

For studying structure of polymer and influence of introduction of additives HZ-1; HZ-2; HZ-4 the method of micro photographing has been applied. Cuts of polymeric fibers photographed at 40 multiple increase, on a microscope "Polar" which had a special nozzle for photographing.

It follows from the analysis of micro photos (fig. 2-3), that hexsaazocyclanes introduction does not break uniformity of the polymeric structure that speaks about good solubility of hexsaazocyclanes in fusion PETP.

In a photo of cross-section cut initial PETP fibres (fig. 2) are observed two areas: amorphous and crystal.

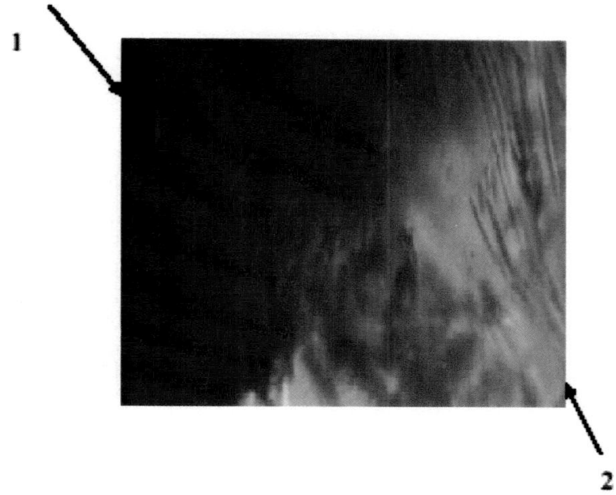

Figure 2. Micro photo of a cross-section cut initial PETP: 1) an amorphous part; 2) a crystal part

In crystal area lamellar and fibrillation formations are visible, but they are not advanced enough. The amorphous part occupies great volume from the general space of polymer.

On fig. 3 micro photos PETP – the fibers modified HZ-1 (a) are represented; HZ-2 (б); HZ-4 (в), and their cross-section cuts. From comparison of a micro photo initial and modified PETP, it is necessary to note, that modified PETP shows the big tendency to crystallization.

In pictures of the fibers painted HZ-1, HZ-2, HZ-4, are visible lamellar and fibrillation formations and are also observed spherallites.

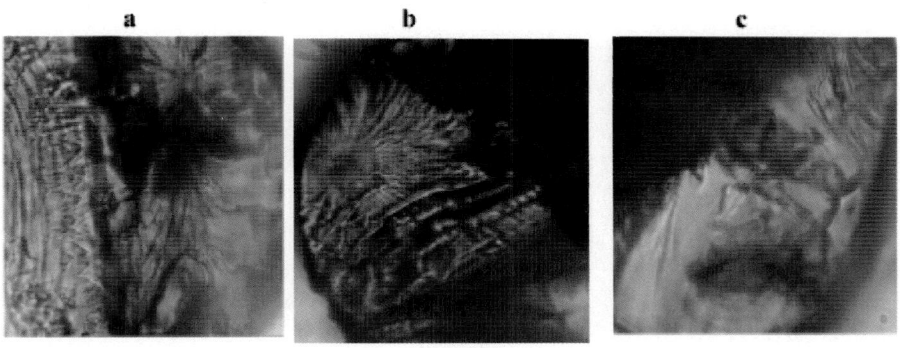

Figure 3. Micro photo of the PETP modified HZ-1 (a); HZ-2 (b) and HZ-3 (c)

It is well visible in micro photos, that the quantity of an amorphous part of the modified fibers decreases. It raises durability of a fibre and improves his mechanical characteristics as the amorphous area is the most vulnerable for action loadings and pressure as it contains areas of the least order in the macromolecules packing, and the places focusing defects.

In micro photos of the modified fibers is the fibrillation structure of a fiber, and visible introduction hexsaazocyclanes in PETP promotes to increase order and density of the structure of the polymer.

It was revealed earlier [2], that hexsaazocyclanes additives cause temperature reduction of polyethylene terephthalate fusion, i.e. they have a plasticization effect on polymer.

Plasticization proves to be true by measurements of electro physical characteristics at modified polyethylene terephthalate (figure 4).

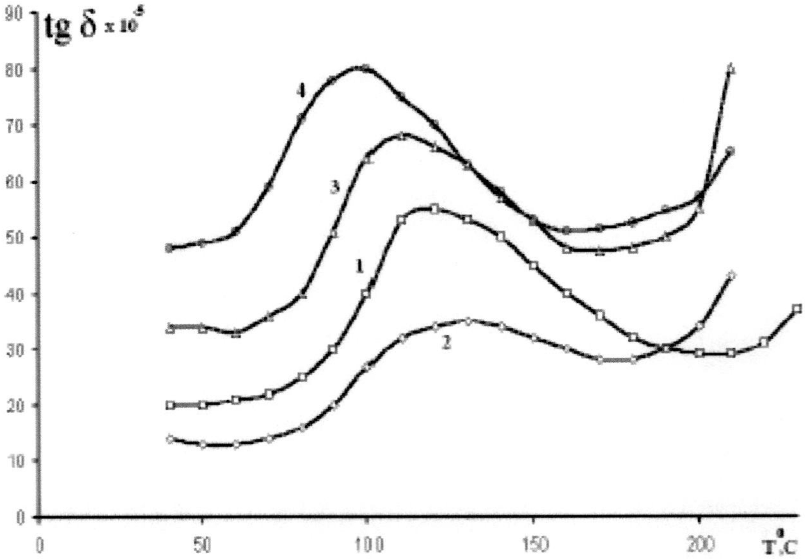

Figure 4. Dependence of a tangent of a corner of dielectric losses on temperature initial and modified samples PETP: 1-initial PETP;2-PETP+0,1% HZ-2;3-PETP+0,5% HZ-2,4-PETP+3,0% HZ-2

For initial polyethylene terephthalate the maximum of a tangent of dielectric losses (tg δ), corresponding depolsegmentation mobility, is observed at 115^0 C.

At concentration of the modifier of 1-3 % sharp increase of value tg δ and displacement of a maximum accordingly on 10^0C and 25^0C in area of lower temperatures that confirms plasticization action of the modifier is observed.

In some works [3, 4] it is supposed, that PETP can have two morphological forms of crystals which define double endothermic effect in the field of fusion. Form I, to which more high-temperature corresponds originally endothermic peak of fusion, has been attributed folded structure, and for the form II responsible for more low-temperature peak of fusion, the crystal structure from more extended circuits has been offered.

Thus the length fold forms I causes a degree of perfection of the crystal form II created by partial expansion fold of form I though the nature remains obscure.

The analysis of micro photos of received samples PETP– fibers has shown, that in the received polymer there are both morphological forms of crystals. Prevalence of one of morphological forms will define, apparently, size of temperature of fusion.

On curves DTA initial PETP - fibres and modified hexsaazocyclanes (fig. 5), the peak of fusion represents the area having one maximum that confirms the assumption made earlier of dependence of temperature of fusion from a morphological structure of polymer.

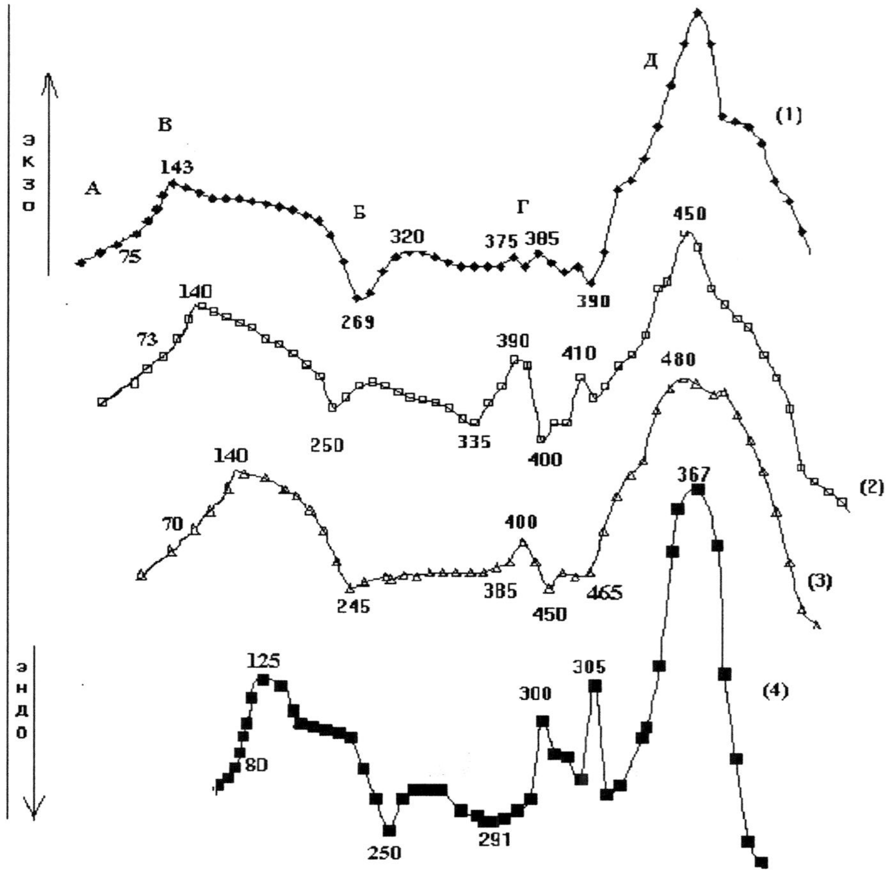

Figure 5. Curves DTA for unmodified (1) and modified HZ-1 (2), HZ-2 (3) and HZ-4 (4) PETP - fibres

Displacement of peak of fusion on curves DTA in area of smaller temperatures specifies that in the modified fibres crystals have mainly morphological form II (the extended circuits of polymers incorporated into crystallites) while in initial PETP crystals mainly have morphological form I (folded structure). Therefore for initial PETP it is observed endothermic effect at temperature 269^{0}C – speaking by fusion flat folded crystallites (morphological form I). At modified PETP – fibers this effect is observed at temperature 245 – 263^{0}C, it speaks fusion of spherallite (the morphological form II).

At heating samples PETP on thermogramms the thermal effects, connected with the fusion crystallites, were fixed. On curves DTA modified PETP – fibers it is visible, that at them the breadth of peak of fusion is more, than at initial PETP.

Proceeding from the aforesaid, it is possible to admit, that modified hexsaazocyclanes PETP - fibres possess the greater degree crystalline state. It once again confirms the conclusion made earlier, that entered molecules hexsaazocyclanes become the additional centers of crystallization, thus, increasing a degree crystallization modified PETP - fibers.

References

[1] Savenkova I.V., Nijazi F.F., Siling S.A., Burykina O.V., *Russian polymer news,* 2001, vol.6, No.4, s. 61-62.

[2] Burykina O.V., Nijazi F.F., Siling S.A., *News high schools Chemistry and chemical technology,* 2002, vol. 45, No. 5, s. 73-74.

[3] Bell P.J., Dumbleton S.H., *Polymer Sci.,* 1969, pt. A-2,V.7,№2, p.1033.

[4] Neaiy D.L., Davis T.G., Kibier C.J., *J. Polymer Sci.,* 1970, pt. A-2, V.8, №2, p. 2141.

In: Order and Disorder in Polymer Reactivity
Editors: G. E. Zaikov and B. A. Howell, pp. 19-27
ISBN 1-60021-145-3

Chapter 3

QUANTUM-CHEMICAL INTERPRETATION OF PEROXIDES DECOMPOSITION

A. Turovskiy[1], I. Golovata[1], E. Zagladko[2], G. Zaikov[2] and O. Romanyuk[1]

[1] Physical-Chemistry and Combustible Minerals Dpt., Pisarshevsky Institute of Physical Chemistry, Ukrainian National Academy of Sciences, 3a, Naukova Str., Lviv 79053, Ukraine

[2] N.Emanuel Institute Biochemical and Physics, Russian Academy of Sciences, 4 Kosygin Street, Moscow 117334, Russia

ABSTRACT

The kinetic parameters of the peroxide's decomposition have been calculated in the PM3 approximation of the semi-empirical method of quantum chemistry. The negative values of the activation entropy for reversible reaction were explained. The linear dependences between thermodynamic parameters of active complex, initial reagents and products of reaction were discovered.

Key words: peroxides destruction, active complex, direct and back reactions, pre-exponential factors, and kinetic and thermodynamic parameters.

Peroxides are widely applied as initiators and photo initiators of radical chain reactions. For conducting of purposeful peroxides synthesizing it is need to know the process kinetic parameters. Experimental investigations are applied with this purpose. Unfortunately, such investigations are not satisfactory in all cases, because peroxides decomposition in many cases is complicated by inductive decomposition. This fact is caused by brut-constants receiving. From the other hand, some troubles of methodological character are arising at the studying of peroxides kinetic of decomposition. Before the synthesis of new series of peroxides it is necessary to lead a preliminary estimation of their kinetic parameters with help of more or less reliable quantum-chemical method.

The aim of this paper is the calculation of kinetic parameters for elementary reactions of carbons and some substituents of peroxides decomposition by PM3 method [1].

As object of study are taking compounds from paper [2]. Numeration of these compounds is the same as in tables and figures.

1. C2H5COOOCOC2H5;
2. nC3H7COOOCOC3H7n;
3. 3.(CH3)2CHCOOOCOCH(CH3)2;
4. nC4H9COOOCOC4H9n;
5. nC5H11COOOCOC5H11n;
6. nC6H13COOOCOC6H13n;
7. nC7H15COOOCOC7H15n;
8. C6H5CH2COOOCOC6H5CH2;
9. C6H5CH=CHCOOOCOC6H5CH=CH;
10. cycloC4H7COOOCOC4H7cyclo;
11. cycloC3H5COOOCOC3H5cyclo;
12. cycloC5H9COOOCOC5H9cyclo;
13. CH3COOOCOC2H5;
14. CH3COOOCOC3H7n;
15. CH3COOOCOC4H9n;
16. CH3COOOCOC6H5;
17. CH3COOOCOC6H4Cl-para;
18. CH3COOOCOC6H4Br-para;
19. CH3COOOCOC6H4CH3-para;
20. CH3COOOCOC6H4OCH3-para;
21. ClCH2COOOCOC6H5;
22. C2H5COOOCOC3H7n;
23. C2H5COOOCOC6H5;
24. nC3H7COOOCOC6H5;
25. C2H5COOCC2H5;
26. (CH3)2CHOOCH(CH3)2;
27. nC4H9OOC4H9n;
28. (CH3)3COOC(CH3)3;
29. (CH3)2(C2H5)COOC(C2H5)(CH3)2;
30. C6H5(CH3)2COOC(CH3)2C6H5;
31. (CH3)3SiOOSi(CH3)3;
32. (CH3)3COOC(CH3)2C2H5;
33. (CH3)3COOC(CH3)2C6H5;
34. (CH3)3COOCH3OCH2OCH2;
35. (CH3)3COOCH2OH;
36. C2H5(CH3)2COOCH2OCOC6H13;
37. C2H5(CH3)2COOC6H13COOCH2;
38. C6H5(CH3)2COO(C6H5)2CCH3;
39. nC4H9OCH2OOC(CH3)2C2H5;

40. C6H10NCH2OOC(CH3)2C2H5;
41. C2H5COOOCOC4H9n;
42. CH3COOOCOC6H4NO2-meta;
43. CH3COOOCOC6H4OCH3-orto;
44. CH3COOOCOC6H4CH3-meta;
45. CH3COOOCOC6H4Br-orto;
46. CH3COOOCOC6H4Br-meta;
47. CH3COOOCOC6H4Cl-orto;
48. CH3COOOCOC6H4Cl-meta;
49. CH3COOOCOCH2CH(CH3)2;
50. CH3COOOCOC6H11cyclo;
51. C6H5COOOCOC6H5;
52. CH3COOOCOCH3.

The following marks are accepted in paper: ΔHr, Sr, ΔFr – thermodynamic parameters of reagents; ΔHc, Sc, ΔFc – enthalpy, entropy and free energy of active complex;

ΔHpr, Spr, ΔFpr – thermodynamic parameters of products; $\Delta H^{\neq}$, $\Delta S^{\neq}$ i $\Delta F^{\neq}$ - activation parameters of reactions. Indexes «p» and «e» by kinetic and thermodynamic characteristics mean calculated and experimental values.

It has been consider, that quite interesting is to find the correlations between thermodynamic characteristics of active complex and initial reagents. Quantum-chemical calculations have been lead for elementary radical monomolecular reactions of destruction a series of peroxide under 363 K degrees.

The ways of decomposition of some alkyl and acyl peroxides are shown on figure 1. Analysis of calculations is shown, that the bond O-O length in active complex increasing on 0,6-0,9 A in comparing with balanced state. For majority of molecules lengthening of O-O bond on average is equal 0,7 A. It allows relate active complexes rather to free active complexes.

Table 1. Values of activation parameters for peroxides termolysis (363K)

№ of compound	$\Delta H^{\neq}$ p kcal/ moll	$\Delta H^{\neq}$ e kcal/ moll	$\Delta F^{\neq}$ p, kcal/ moll	$\Delta S^{\neq}$ p, cal/moll/K	$\Delta S^{\neq}$ e, cal/moll/K	Lg Ap	Lg Ae
1	29,90	30,02	26,03	10,66	8,28	15,36	14,81
3	23,07	26,02	19,86	8,83	7,30	14,94	14,58
10	15,64	22,92	17,78	-5,89	-7,14	11,50	11,21
15	27,06	31,82	21,51	15,30	12,95	16,44	15,90
16	30,34	34,52	27,64	7,43	5,58	14,61	14,18
17	36,80	31,92	34,14	7,33	5,28	14,59	14,10
25	40,56	40,52	36,44	11,35	9,27	15,53	15,04
28	25,31	29,62	26,49	-3,26	-3,41	12,12	12,08
31	26,03	28,02	26,20	-0,46	-0,03	12,77	12,87
32	25,04	27,72	24,54	1,37	1,17	13,20	13,15
51	29,98	31,32	25,59	12,10	12,09	15,7	15,70
52	27,34	29,32	25,48	5,13	5,58	14,07	14,18

It is given from fig.1, that an activation free energy is depended from free energy of initial reagent creation. Thus, bigger reaction ability is agreed with less value of potential border and with less reaction ability of reaction products. There is linear dependence between active complex free energy and initial reagent free energy. Similar dependence should be investigated between products free energy and active complex free energy (fig.2).

Calculations of kinetic and thermodynamic parameters are shown in table 1.

It is given from table 1, that an activation free energies, activation enthalpies, activation entropies and pre-exponential factors are satisfactory compared with experimental ones. Complex activation entropy is increasing, but that factor is not contradicted with well-known ideas. Changing is occurred with increasing of vibration part of active complex entropy. Rotation part of active complex entropy is changed lightly (table 2).

Table 2. Entropy and its parts (T=363 K)

№ of compound	S, cal/moll/K		Svib, cal/moll/K		Srot, cal/moll/K	
	Active complex	Initial reagent	Active complex	Initial reagent	Active complex	Initial reagent
1	127,88	117,22	54,67	45,23	31,40	30,88
3	145,94	137,11	64,52	61,87	31,52	31,91
10	126,97	132,86	60,13	57,340	32,13	32,81
15	140,34	125,04	55,29	51,12	31,14	31,84
16	102,95	95,52	45,30	44,93	32,07	32,07
17	129,83	122,5	51,56	50,01	33,28	34,02
25	153,56	142,21	30,64	27,73	27,30	27,42
28	126,05	129,31	57,06	50,48	30,97	30,61
31	125,79	126,25	70,08	68,47	31,30	31,33
32	120,80	119,43	47,18	45,60	31,13	31,23
51	114,97	102,87	55,18	52,84	31,13	32,07
52	142,00	136,87	44,09	50,18	29,71	30,02

Negative values of activation entropy for some acyl peroxides can be explained by increasing the interaction between valence non-combined atoms of oxygen and α-oxygen of peroxide bridge. For example for peroxide №10 energy of valence non-combined atoms is equal 0,05 eV in active complex and for molecule with positive activation entropy (№1) - 0,01 eV. For molecules of alkyl peroxides with tert-butyl substituent (№28) and for siliceous-organic compound (№31) activation entropy is negative too. It may be to cause with stronger interaction between tert-butyl carbon atom and α-oxygen of peroxide bridge in active complex. Those values for such peroxides are equal 0,09 eV and 0,43 eV accordingly.

Interactions of valency non-combined atoms for aforesaid peroxides are lead to formation thee-member cycles. Such cycles are tougher, than in initial state. Such cycles have higher frequency of extra-plain deformations of cycle. Such frequency is called as pseudo-rotation. Decreasing of entropy for such type of compounds is caused by this mode. For all another peroxides with positive values of activation entropy similar phenomenon is not investigated.

Starting from, that

$$\Delta S_p = \Delta S^{\neq}_{np} - \Delta S^{\neq}_{зв} \text{ and } \Delta F_p = \Delta F^{\neq}_{np} - \Delta F^{\neq}_{зв} \quad (1)$$

kinetic parameters for back reactions are estimated (see table.3).

Table 3. Values of kinetic and thermodynamic reactions for back reactions (T=363 K)

№ of compound	$\Delta H^{\neq}$ p, kcal/ moll	$\Delta F^{\neq}$ p, kcal/ moll	$\Delta S^{\neq}$ p., cal/moll/K	Lg Areaction (direct)	Lg Areaction (back)	Lg Aexperim. (direct reaction)
1	5,78	-146,45	-2,7	14,18	10,60	14,81
3	6,38	-171,22	-4,11	13,62	10,02	14,58
10	2,34	-129,92	-1,83	10,78	10,85	11,21
15	7,16	-161,90	-1,95	15,44	11,42	15,90
16	6,15	-103,89	-1,25	12,86	10,58	14,18
17	2,94	-103,93	-0,16	12,83	12,84	14,10
25	4,36	-51,33	-1,62	13,80	11,50	15,04
28	7,09	-79,04	-3,25	13,35	10,12	12,08
31	2,36	-243,08	-1,84	14,61	10,56	12,87
32	4,08	-87,16	-1,86	11,65	11,41	13,15
51	6,93	-71,09	-2,42	13,61	12,31	15,70
52	5,76	-135,55	-2,39	15,14	10,72	14,18

Activation enthalpy for back reactions is in interval 2-7 kcal/moll.

Thus under doubt is correlation, that activation energy of monomolecular decomposition (for this series) is equal with heat of reaction.

Pre-exponential factors of back reactions are within reasonable limits under 363 K degrees. Pre-exponential factors of direct reaction, which have been estimated from such factors, are satisfactory sorted with experimental data of ones.

Dependence between active complex free energy and initial reagents free energy for those series of compounds is shown in fig.1 (б).

This dependence is calculated from experimental values of $\Delta F^{\neq}$ and theoretical values of reagents ΔF. This relation can be showed by equation:

$$\Delta F_c = (1{,}01 \pm 0{,}01)\Delta F_r + (30{,}14 \pm 1{,}10)$$

R=0,998 SD=2,78 (2)

Dependence between active complex free energy and initial reagents free energy, which are calculated theoretically, is shown in fig.2 (a).

Empirical equation looks as:

$$\Delta F_c = (1{,}25 \pm 0{,}08)\Delta F_r + (54{,}17 \pm 11{,}83)$$

R=0,98 SD=11,41 (3)

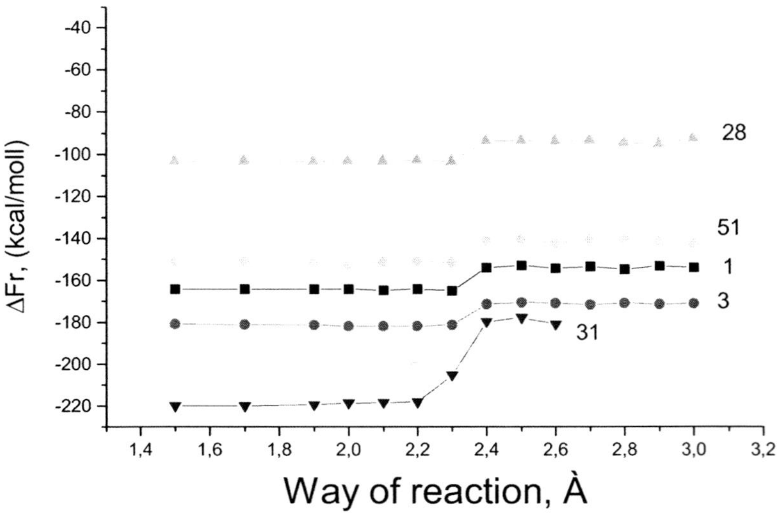

a)

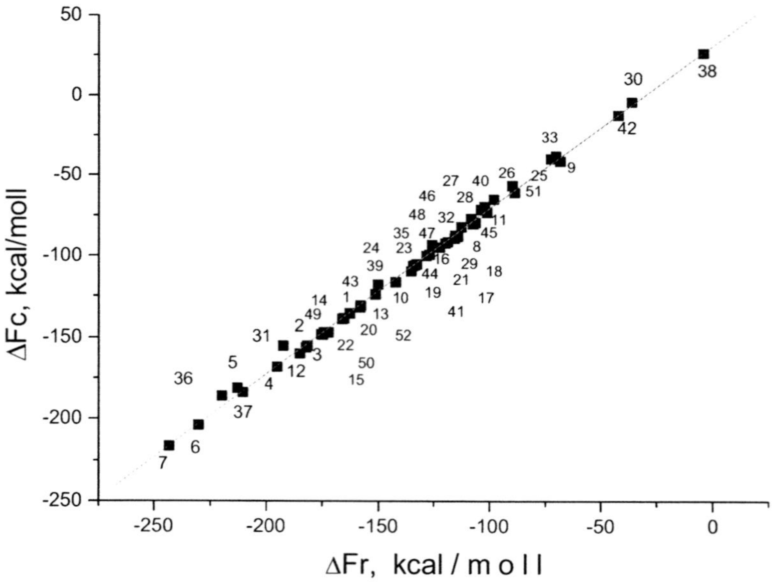

b)

Figure 1. Ways of decomposition for some peroxides and (a) dependence between complex free energy and initial reagents free energy for those compounds (b).

Dependence between active complex free energy and products free energy, which are calculated theoretically, is shown in fig.2 (b):

$$\Delta F_c = (0{,}99 \pm 0{,}01)\Delta F_{np} + (6{,}93 \pm 1{,}88)$$

R=0,99 SD=2,47 (5)

Dependence between initial reagents free energy and products free energy, which are calculated theoretically, is shown in fig.2 (c). Empirical equation looks as:

$$\Delta F_c = (1{,}25 \pm 0{,}08)\Delta F_{np} + (47{,}81 \pm 10{,}88)$$

R=0,98 SD=10,49 (6)

Value $\Delta H^{\neq}$ for destruction of compounds has been calculated from equation:

$\Delta H^{\neq}$=E - RT, (7)

Where T=900 K; R- universal gas constant; E-activation energy of reaction.

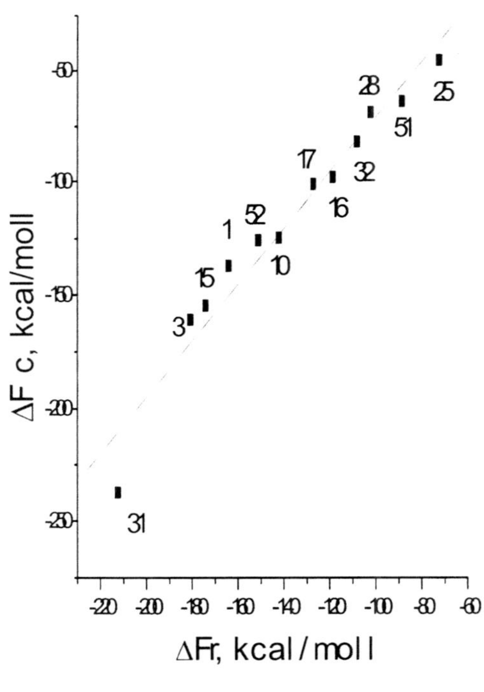

a)

Figure 2 continued on next page

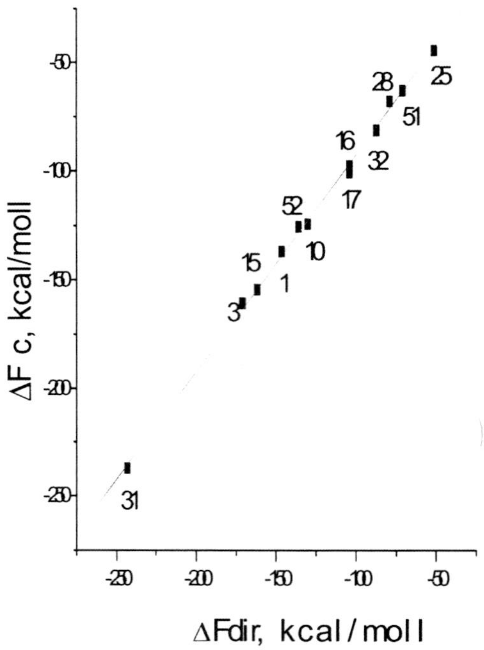

b)

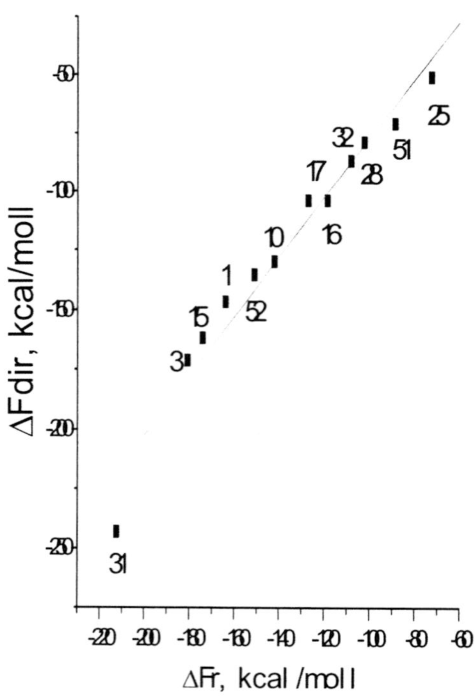

c)

Figure 2. Theoretical dependence of active complex free energy and initial reagents free energy (a). Dependence between active complex free energy and products free energy (b). Dependence benween products free energy and initial reagents free energy (c).

Value $\Delta S^{\neq}$ have been calculated from data for lg A according to equation:

$$A = \chi e \cdot \frac{kT}{h} \cdot e^{\frac{\Delta S^{\neq}}{R}} \quad (8)$$

Where χ-transmission coefficient, χ=1; k- Boltsman's constant;

Plank constant; $\Delta S^{\neq}$ - activation entropy.

REFERENCES

[1] M.W. Schmidt, K.K. Baldridge, J. A. B o a t s. *J.Comput.Chem.,* 11, 1347 (1993)[1]

[2] Denisov E. T. Homolytic liguid-phase reactions constants of speed. Moscow, *"Nauka",* 1971, page 711.

In: Order and Disorder in Polymer Reactivity
Editors: G. E. Zaikov and B. A. Howell, pp. 29-36
ISBN 1-60021-145-3

Chapter 4

QUANTUM-CHEMICAL INTERPRETATION OF CARBON PYROLYSIS KINETICS

A. Turovskiy[1], I. Golovata[1], E. Zagladko[2], G. Zaikov[2] and O. Romanyuk[1]

1 Physical-Chemistry and Combustible Minerals Dpt., Pisarshevsky Institute of Physical Chemistry, Ukrainian National Academy of Sciences, 3a, Naukova Str., Lviv 79053, Ukraine

2 N.Emanuel Institute Biochemical and Physics, Russian Academy of Sciences, 4 Kosygin Street, Moscow 117334, Russia

ABSTRACT

The kinetic parameters of the hydrocarbon's pyrolysis have been calculated in the PM3 approximation of the semi-empirical method of quantum chemistry. The ways of some compounds destruction were calculated. The linear dependences between free energies of active complex, initial reagents and products of reaction were discovered. It was shown that high and low values of pre-exponential factors caused by vibration part of activation entropy.

Key words: active complex, carbons, pyrolysis, direct and back reactions.

INTRODUCTION

Constants of speed and balance are the most important characteristics of any chemical system. With some other parameters (such as conversion, heat- and mass-transmission coefficients) they allows us to calculate parameters of technological process (pressure, temperature, catalyst activity etc.).

In study of radical reactions there are some troubles, because creation of simple and valid methods for calculation of speed constants for different reactions is an actual problem.

There are some methods of kinetic parameters estimation for elementary reactions. The most well known is the method with phenomenological character. On the basis of experimental data and thermodynamic idea are drown the correlation equations. Such equations are described the reaction possibility of molecules and radicals for particular series.

One of the trends for calculation of electronic structure of compound models and radicals and ways of reaction is the quantum-chemical method of calculations. Creation of more perfect semi-empirical methods in quantum chemistry with applying of computers make this trend more long-term in studying of kinetics and mechanisms of different elementary reactions.

The aim of this paper is the calculation of kinetic parameters for elementary reactions of carbons and some substituents of carbons pyrolysis by PM3 method [7].

We are considering the quite interesting is search of correlations between thermodynamic characteristics of active complex and initial reagents.

As objects of study are compounds from [8] and [9]. Numeration of these compounds is same in tables and figures.

1. CH4;
2. C6H5CH2-H;
3. C2H6;
4. C2H5-CH3;
5. 5.n-C3H7-CH3;
6. iso-C4H10;
7. C (CH3)3-CH3;
8. C5H12;
9. C6H14;
10. C6H14;
11. C7H16;
12. CH3CH2-C(CH3)2CH2CH3;
13. C4H9-C4H9;
14. C7H15-CH3;
15. CH3CH=CH2;
16. C4H6;
17. CH2=CHCH2C(CH3)3;
18. CH2=CH(CH2)2CH=CH2;
19. CH3-CH2CN;
20. C2H5I;
21. C6H5CH2-CH3;
22. (CH3)3C-C(CH3)3;
23. (CH3)3CCH(CH3)2;
24. (CH3)3CCN; 25. C3H6.

In paper are accepted such marks: ΔHr, Sr, ΔFr – thermodynamic parameters of reagents; ΔHc, Sc, ΔFc – enthalpy, entropy and free energy of active complex;

ΔHpr, Spr, ΔFpr – thermodynamic parameters of products; $\Delta H^{\neq}$, $\Delta S^{\neq}$ i $\Delta F^{\neq}$ - activation parameters of reactions. Indexes «p» and «e» by kinetic and thermodynamic characteristics mean calculated and experimental values.

Quantum-chemical calculation have been conducted for elementary radical monomolecular reactions of a number of carbons destruction under 900 K degrees.

The ways of destruction some carbons on free radicals are demonstrated on figure 1.

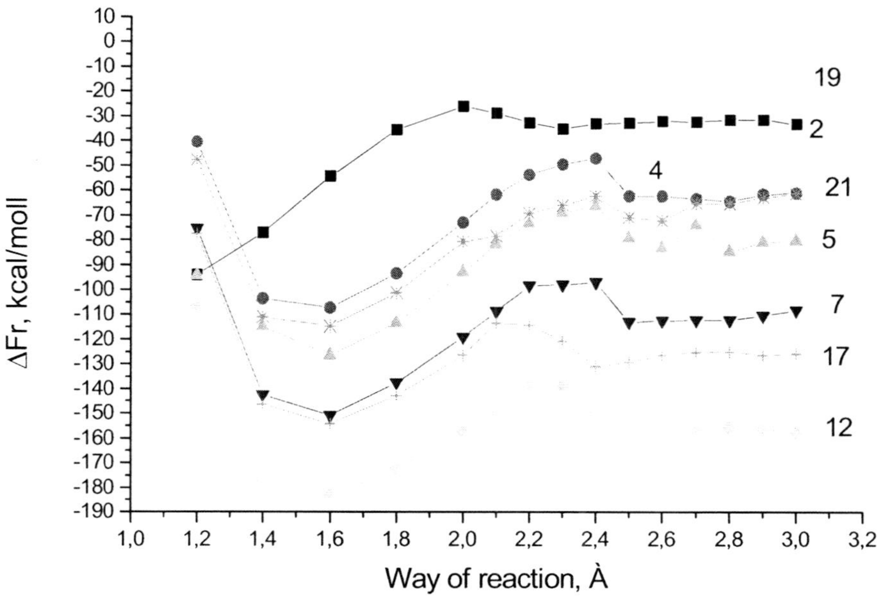

Figure 1. Decomposition of some carbons (900 K).

According to calculations the length of bond in active complex is more than 0,7-1,3 A in comparing with the value for balance state. It is 45-85% of balanced state bond length. For majority of molecules growth of length is in average 1A, because those active complexes are free active complexes. But according to [8], for group of free active complexes are concerned complexes with length of C-C bond (approximately 3,5-4A). Comparing our calculations with experimental data is shown, that the length of bond C-C in active complex is a little less as in literature [8]. In dependence of substituent the length of neighbour atoms by C-C are changed not more 0,05 A.

Results of kinetic and thermodynamic parameters calculations for decomposition of some molecules are given in table 1.

Calculated values of free activation energy, activation enthalpy, activation entropy and pre-exponents are satisfactory compared with experimental ones.

It is shown in table 2 that activation entropy of complex is slightly increased. It is not object well-known ideas. Change is occurred through increasing of vibration part of active complex entropy. Rotation part of active complex entropy is changed insignificantly.

Table 1. Values of kinetic and thermodynamic parameters of carbons pyrolysis (900 K)

Type of compound	$\Delta H^{\neq}$ p ccal/ moll	$\Delta H^{\neq}$ e ccal/ moll	$\Delta F^{\neq}$ p, ccal/ moll	$\Delta F^{\neq}$ e ccal/ moll	Lg Ap	Lg Ae
2. C6H5CH3	74.21	75.43	73.18	74.17	13.54	13.60
4. C3H8	87.08	80.30	70.75	63.65	17.51	17.60
5. n-C4H10	84.13	84.60	61.13	64.17	19.24	18.58
7. (CH3)4C	88.26	81.70	70.80	65.77	17.80	17.40
12.C2H5C(CH3)2C2H5	79.17	75.50	66.14	62.81	16.65	16.57
13. C8H18	61.14	67.00	52.55	55.3	15.50	16.30
14. C8H18	67.32	74.60	56.3	64.07	16.13	16.00
17.CH2=CHCH2C(CH3)3	62.69	63.80	52.98	54.17	15.79	15.78
19. CH3CH2CN	66.39	71.00	64.24	67.85	13.83	14.10
21. C6H5C2H5	71.22	68.40	66.64	63.27	14.46	14.60
25. C3H6	93.46	92.30	76.68	74.84	17.63	17.80

Table 2. Values activation entropy and pre-exponents for reaction of carbon pyrolysis (900K)

Type of compound	$\Delta S^{\neq}$ p, cal/moll* K	$\Delta S^{\neq}$ e, cal/moll* K	Svib, cal/moll*K		Srot, cal/moll*K	
			Active complex	Initial reagent	Active complex	Initial reagent
2. C6H5CH3	1.14	1.40	49.52	48.39	30.25	30.25
4. C3H8	18.15	18.50	43.69	26.65	27.04	25.94
5. n-C4H10	25.56	22.70	64.20	39.62	28.76	27.78
7. (CH3)4C	19.40	17.70	73.79	54.94	29.57	28.58
12.C2H5C(CH3)2C2H5	14.48	14.10	94.24	80.2	31.55	31.10
13. C8H18	9.54	13.00	105.10	95.37	32.5	32.49
14. C8H18	12.25	11.70	107.82	95.37	32.49	32.49
17.CH2=CHCH2C(CH3)3	10.79	10.70	89.81	79.26	31.41	31.18
19. CH3CH2CN	2.39	3.50	27.67	25.15	27.01	27.14
21. C6H5C2H5	5.09	5.70	67.64	62.56	31.34	31.34
25. C3H6	18.65	19.40	39.88	22.91	26.98	25.30

The wide space of pre-exponential factors for investigated carbons (13,6-18,6) can be explained by part of deformation vibrations. High increasing of bond in active complex and insignificant changes neighbour with reaction centre bonds allows us to suppose, that only cause change in vibration part of activation entropy must be change of deformation vibrations in active complex.

Kinetic parameters of back reactions have been estimated, coming from that

$$\Delta S_p = \Delta S^{\neq}_{np} - \Delta S^{\neq}_{зв} \text{ and } \Delta F_p = \Delta F^{\neq}_{np} - \Delta F^{\neq}_{зв} \quad (1)$$

Activation enthalpy for back reactions is in space 0,5-5 kcal/moll. In literature is considered speed constants are equal pre-exponential factor. This supposition is true, but not for all cases.

Thus, in doubt is idea that activation energy for compounds monomolecular destruction is equal with the heat of reaction.

Pre-exponential factors of back reactions under 900 K are quite true, because estimated values of pre-exponential factors for main reaction satisfactory compared with experimental data.

Let to consider in more details reaction of recombination. Active complex is same for direct and back reactions, because thermodynamic characteristics of active complex (Δ Hc=71,74 kcal/moll, Sc=100,34 cal/(moll*K)) and calculated thermodynamic parameters for free radical ($\Delta H_{CH_3^\bullet}$ =34,85 kcal/moll, $S_{CH_3^\bullet}$ = 60,46 cal/(moll*K)), and as result activation enthalpy of recombination $\Delta H^{\neq}_{PEK}$ =2,04 kcal/moll, activation entropy $\Delta S^{\neq}_{PEK}$ =-20,58 cal/(moll*K).

According to [8] $\Delta S^{\neq}_{PEK}$ =-25,66 cal /(moll*K). Author is allowed, that k=2,3·10^{11} l/(moll*sec). According to our calculations k=4,3·10^{10} l/(moll*sec). The best compliance according to [9] for recombination of radicals under 900 K k=(2±0,15)·$10^{9}T^{1/2}$ = (6±0,45)·10^{10} l/(moll*sec). Thus, our calculations are better correlated with experimental data.

Even increasing length of C-C bond in active complex for reactions of carbons recombination not a cause of groups free rotation around C-C bond (limits in active complex are equal (0,5-1 kcal/moll), because under doubt is creation of free active station. This admission is confirmed that vibration part of active complex activation entropy for decomposition of C_2H_6 with calculation for one CH3 group is equal 19,8 cal/(moll*K) and S_{vib} =3,75 cal/(moll*K) for methyl radical. It talks about quite essential interaction of methyl groups in active complex. It is under doubt, that this interaction can be modulated as two free radicals.

Taking into account, that under 900 K factor for the methyl radicals z=4·10^{11} l/(moll*sec) [8], then s-recombination factor for methyl radicals s=1,08·10^{-1}. This value in [8] is equal 6,3·10^{-2}. It seems to be probably, that our calculations for this value are better.

According to figure 1, activation free energy depends from the value of free energy of free radical creation (increasing of reagent free energy lead to increasing of complex free energy). Because increasing of activation free energy caused decreasing of speed constant. Thus, can to suppose of linear dependence between active complex free energy and reagents free energy. Similar dependence should be investigated between product free energy and active complex free energy.

Dependence between free energy of active complex and free energy of initial reagents, which calculated theoretically, are shown on figure 2 (a).

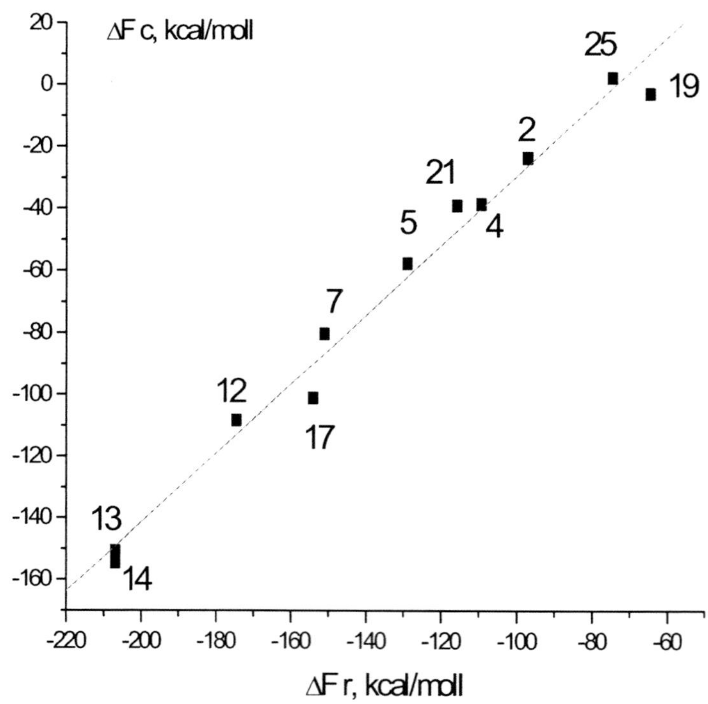

a)

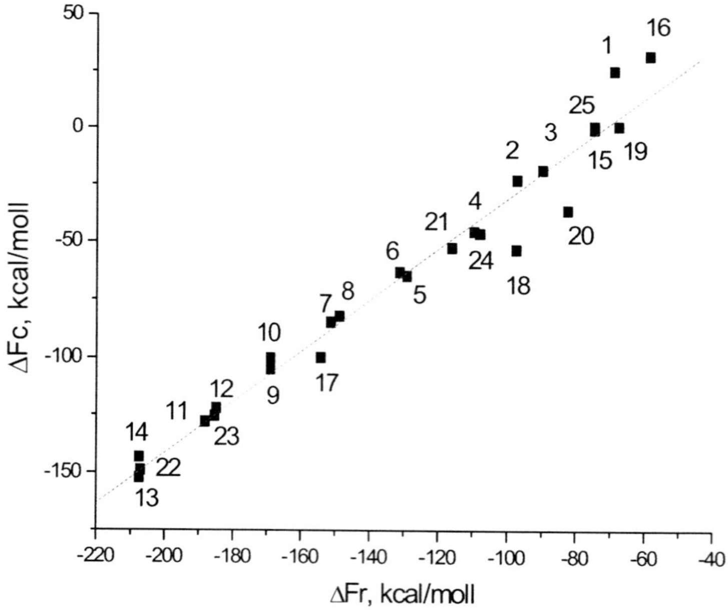

b)

Figure 2. Theoretical (a) and experimental (b) dependence of active complex free energy from reagents free energy (900 K).

Empirical equation can be shown:

$$\Delta F_c = (1{,}12 \pm 0{,}05)\Delta F_r + (82{,}02 \pm 6{,}86)$$

R=0,992 SD=7,40 (2)

Dependence between free energy of active complex and free energy of reagents for this number of compounds is shown on figure 2 (b). Those data are calculated from experimental data $\Delta F^{\neq}$ and theoretical values ΔF of reagents. Such dependence can be shown by equation:

$$\Delta F_c = (1{,}098 \pm 0{,}04)\Delta F_r + (78{,}32 \pm 5{,}89)$$

R=0,983 SD=10,13 (3)

Value $\Delta H^{\neq}$ for destruction of compounds has been calculated from equation:

$$\Delta H^{\neq} = E - RT, \quad (4)$$

Where T=900 K; R- universal gas constant; E-activation energy of reaction.

Value $\Delta S^{\neq}$ have been calculated from data for lg A according to equation:

$$A = \chi e \cdot \frac{kT}{h} \cdot e^{\frac{\Delta S^{\neq}}{R}} \quad (5)$$

where χ-transmission coefficient, χ=1; k- Boltsman's constant;

Plank constant; $\Delta S^{\neq}$ - activation entropy.

BIBLIOGRAPHY

[1] Hammet L. *Osnovy phizicheskoy himii.* Translation from English under L.S.Efros redaction. Moscow, "Mir", 1972, 534 pages.

[2] Leffler J.E., Grunwald R., *Rates and Equilibria of Organic Reactons.* N.Y., 1963. 458 p.

[3] Zhdanov Yu.A., Minkin V.I. *Korelyatsionny analiz v organicheskoy himii.* Rostov-na-Donu, Publishing house RGU, 1966. 470 pages.

[4] Palm V.A. *Osnovy kolichestvennoy teorii organicheskih reaktsii.* Leningrad, "Himiya", 1967. 356 pages.

[5] Bagdasaryan H.S. *Teoriya radikalnoy polimerizatsii.* Moscow, Publisher house of USSR Academy of Sciences, 1966. 300 pages.

[6] Bazulevskiy M.V. *Metod MO I reactsionnnaya sposobnost` organicheskih molekul.* Moscow, "Himiya", 1969. 303 pages.

[7] M.W. Schmidt, K.K. Baldridge, J.A.Boats. *J.Comput.Chem.,* 11, 1347 (1993)[1]

[8] Stepukhovich A.D., Ulitskiy V.A. *Kinetica I termodinamica radicalnuh reactsii krekinga.* Moscow, "Himiya", 1969. 303 pages.

[9] Vedeneev V.I., Kibkalo A.A. *Konstanty skorocti gazofaznyh reactsiy.* Moscow, "Nauka", 1972. 164 pages.

In: Order and Disorder in Polymer Reactivity
Editors: G. E. Zaikov and B. A. Howell, pp. 37-51
ISBN 1-60021-145-3

Chapter 5

KINETICS OF ACTIVATED BY Et_4NBrA-OXYCYCLOHEXYL –PEROXIDES DECOMPOSITION. SUPRAMOLECULAR MODEL

M. A. Turovskyj, I. O. Opeida[a], O. M. Turovskaya, O. V. Raksha, N. O. Kuznetsova and G. E. Zaikov[b]
Department of Chemistry, Donetsk National University,
Universitetskaja 24, 83055 Donetsk, Ukraine, E-mail: turov@dongu.donetsk.ua
[a]Institute of Physico-Organic and Coal Chemistry of the National Academy of Sciences of Ukraine, R. Luxemburg 70, 83114 Donetsk, Ukraine
[b]Institute of Biochemical Physics, Russian Academy of Sciences, Kosygina 4, 117334 Moscow, Russian Federation, e-mail: chembio@sky.chph.ras.ru

ABSTRACT

Cyclohexanone peroxides decomposition in the presence of tetraethylammonium bromide in acetonitrile has been studied. The onium salt shows catalytic properties in the system. The scheme of the peroxides activation and decomposition in the presence of quaternary onium salt is proposed. It is substantiated by kinetic methods as well as by molecular modeling methods. It has been shown that peroxides decomposition in the presence of tetraethylammonium bromide proceeded according to supramolecular mechanism. Cyclohexanone peroxides in the presence of Et_4NBr effectively initiate the radical chain cumene oxidation.

Key words: oxycycloalkylperoxides, quaternary ammonium salts, catalysis, supramolecular mechanism, initiation efficiency, molecular modeling

INTRODUCTION

In radical chain reactions the initiation stage determines the rate of process proceeding and thus the process efficiency and their technological parameters as a whole. Peroxides thermal instability and significant dependence of thermolysis rate upon peroxides structure allow using peroxide compounds as initiators in wide temperature interval from 40 to 200^{0}C [1, 2]. Chemically activated peroxides decomposition including peroxide decomposition with tertial amines and onium salts assistance provides radical chain reactions initiation even at 20 - 30 ^{0}C [3-6]. To choose the most effective peroxide initiator it is necessary the following: 1)to create the standartised model technics which provide the comparable characteristics obtaining in the peroxide initiators series; 2)screening realization; 3)establishment of relationship between peroxide structure and their initiation activity.

Kinetics and mechanism of chemically activated peroxides decomposition in solution especially activated by onium salts are very sensitive to the medium properties and initial reactants concentration. Thus keeping to the equal conditions for decomposition and initiation reactions proceeding is of grate importance for radical forming estimation at peroxide decomposition investigation in solution.

Previous investigations have shown the quaternary ammonium salts was able to accelerate the benzoyl peroxide decomposition on -O-O- bond as compared with its thermolysis on several orders [4-10]. Benzoyl peroxide activated by onium salts effectively initiates the methylmetacrylate and acrilonitryle low-temperature polymerization [11, 12], liquid phase hydrocarbons oxidation [13-16]. Results obtained by varying of ammonium activator chemical structure and solvent nature have shown the acceleration effect in benzoyl peroxide decomposition reaction and radicals yield were determined by electrodonating properties of quaternary ammonium salts anion [9]. But it have been established also the benzoyl peroxide decomposition rate depended on the ammonium salt cation too. In particular, activating action of tetraethylammonium chloride (Et_4NCl) at interaction with benzoyl peroxide in acetonitrile is two orders higher as compared with triethylhydrochloride ($[Et_3NH]Cl$) [8].

The present work is devoted to the investigation of cycloaliphatic ketones peroxides (I – α-oxycyclohexylhydroperoxide, II – α,α′-dioxydicyclohexylperoxide) interaction with tetraethylammonium bromide and their initiating activity in the reaction of low-temperature liquide-phase radical chain cumene oxidation.

EXPERIMENTAL

The peroxides purity (98.9 %) was controlled by iodometry method. Tetraethylammonium bromide was recrystallized from the saturated acetonitrile solution by addition of diethyl ether abundance. The salt purity (99.6 %) was determined by argentummetric titration with potenciometric fixation of the equivalent point. Acetonitrile was purified according to [17]. Its purity was controlled by electroconductivity χ value, which was within $(8.5\pm0.2)\cdot10^{-6}$ $Om^{-1}\cdot sm^{-1}$ at 303K. Cumene was subjected to acid - alkali purification with subsequent desiccation under calcium chloride and distillation according to [18]. Reactions of cyclohexanone peroxides decomposition were carried out in the glass soldered ampoules in argon atmosphere. To control the proceeding of peroxides thermolysis and their decomposition in the presence of tetraethylammonium bromide the iodometric titration with potentiometric fixation of the equivalent point was used. Cumene oxidation was carried out in acetonitrile medium in the glass reactor placed into thermostat. The reactor was shaking up with a frequency which provided mixing sufficient for the reaction proceeding in the kinetic region. Accuracy of the determination of the absorbed oxygen amount is 5%.

Cumene Oxidation Initiation by Oxycyclohexyl Peroxides in the Presence of Tetraethylammonium Bromide

Cyclohexanone peroxide compounds (R^1OOR^2) in the presence of tetraethylammonium bromide initiate the cumene oxidation reaction even at 308 - 340K (Fig. 1). Tetraethylammonium bromide as well as peroxide compounds I and II don't initiate cumene oxidation process at mentioned temperature interval.

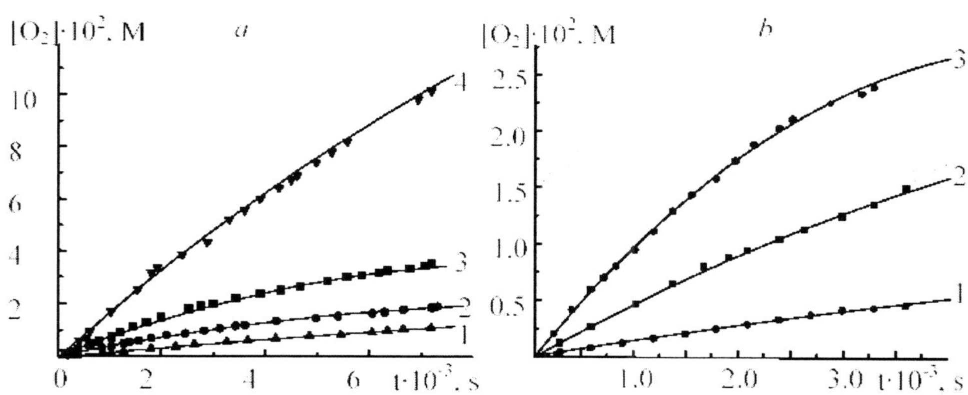

Figure 1. Kinetics of oxygen consumption in cumene oxidation initiated by α-oxycyclohexylhydroperoxide (*a*) and α,α'-dioxydicyclohexylperoxide (*b*) with $Et_4N^+Br^-$ assistance in acetonitrile (*a*: 1 –308 K, 2 – 318 K, 3 – 328 K, 4 - 340 K; *b*: 1 – 313 K, 2 – 328 K, 3 – 338 K; [Cum] = 3.58 M, $[R^1OOR^2] = [Et_4N^+Br^-] = 2\cdot10^{-2}$ M, acetonitrile – cumene (1:1) medium).

Maximum initiated oxidation rates are observed in the initial reaction period. During first 15 min kinetic curves of oxygen consumption are linear. The initial oxidation rate was determined from this period. The corresponding values are presented below:

α – oxycyclohexylhydroperoxide				
T, K	308	318	328	340
$W_{ox}\cdot10^6$, mol/(dm^3·s)	1.9	4.1	7.9	18
α,α' – dioxydicyclohexylperoxide				
T, K	313	328	338	
$W_{ox}\cdot10^6$, mol/(dm^3·s)	1.35	4.5	9.3	

Initiation rate (W_i) was calculated from equation (1) where W_{ox} –cumene oxidation initial rate, k_p, k_t – rate constants of chain propagation and termination correspondingly.

$$W_i = \frac{W_{ox}^2}{\left(\frac{k_p}{\sqrt{2k_t}}\right)^2 [RH]^2} \quad (1)$$

Temperature dependence of $k_p / \sqrt{2k_t}$ was determined within 303 – 339 K temperature interval at cumene oxidation in acetonitrile medium (cumene – acetonitryle 1:1) initiated by benzoyl peroxide in the presence of quaternary ammonium halides and by AIBN. It is described by equation (2).

$$\ln(k_p/\sqrt{2k_t}) = (5.1 \pm 1.4) - (3.5 \pm 0.5)\cdot T^{-1} \quad (2)$$

Initiation rate values for liquid phase radical chain cumene oxidation initiated by peroxide compounds I and II in the presence of $Et_4N^+Br^-$ are presented below.

α – oxycyclohexylhydroperoxide				
T, K	308	318	328	340
$W_i\cdot 10^7$, mol/(dm^3·s)	0.91	2.07	3.91	9.50
α,α′ – dioxydicyclohexylperoxide				
T, K	313	328	338	
$W_i\cdot 10^7$, mol/(dm^3·s)	0.32	1.27	2.87	

α – oxycyclohexylhydroperoxide and α,α′ – dioxydicyclohexylperoxide decomposition activated by tetraethylammonium bromide have been investigated in acetonitrile (AN) as well as in acetonitrile – cumene mixture (AN – Cum) with 1:1 components volume ratio. Kinetic parameters of peroxide compounds I and II decomposition activated by $Et_4N^+Br^-$ in the equal peroxide and ammonium salt concentration conditions ($[R^1OOR^2]_0 = [Et_4N^+Br^-]_0 = 2\cdot 10^{-2}$ M) appeared to be sensitive to medium properties:

α – oxycyclohexylhydroperoxide					
T, K		323	333	343	353
k·10^3, dm^3/(mol·s)	AN	1.20 ± 0.04	2.6 ± 0.1	5.3 ± 0.2	10.0 ± 0.4
	AN-Cum	0.67 ± 0.02	1.72 ± 0.04	3.5 ± 0.1	6.3 ± 0.3
α,α′ – dioxydicyclohexylperoxide					
T, K		323	333	343	353
k·10^3, dm^3/(mol·s)	AN	0.34 ± 0.02	0.80 ± 0.04	1.6 ± 0.1	3.7 ± 0.2
	AN-Cum	0.20 ± 0.01	0.51 ± 0.03	1.23 ± 0.05	2.31 ± 0.08

Temperature dependences of the second order rate constants of α – oxycyclohexylhydroperoxide and α,α′ – dioxydicyclohexylperoxide decomposition activated by $Et_4N^+Br^-$ in acetonitrile and cumene – acetonitrile mixture are described by the following equations:

R^1OOR^2	Solvent	k, $dm^3/(mol \cdot s)$	
I	AN	$(8.71 \pm 0.08) \cdot 10^7 \exp((-67000 \pm 2000)/RT)$	(3)
	AN – Cum	$(2.02 \pm 0.07) \cdot 10^8 \exp((-71000 \pm 3000)/RT)$	(4)
II	AN	$(3.98 \pm 0.09) \cdot 10^8 \exp((-75000 \pm 3000)/RT)$	(5)
	AN – Cum	$(7.95 \pm 0.06) \cdot 10^8 \exp((-78000 \pm 3000)/RT)$	(6)

Rates of the peroxide decomposition (W) activated by $Et_4N^+Br^-$ at temperatures of initiated cumene oxidation (Table 1) were calculated from the equation (7) where rate constants of the peroxide decomposition (k) under considered conditions were estimated from dependences (3-6).

$$W = k[ROOR][QX] \qquad (7)$$

Initiation efficiency (e) was estimated with taking into account of peroxide catalytic decomposition with two radicals formation accordingly to the expression (8):

$$e = \frac{W_i}{2W} \qquad (8)$$

One can see that free radicals yields under cyclohexanone peroxides decomposition activated by tetraethylammonium bromide are less then 1. This fact can be explained by simultaneous proceeding of some rearrangements and decomposition with molecular products formation.

Table 1. Kinetic parameters of peroxides I and II decomposition with $Et_4N^+Br^-$ assistance and cumene oxidation reaction initiated by them

T, K	AN-Cum			AN	
	$W_i \cdot 10^7$, $mol/(dm^3 \cdot s)$	$W \cdot 10^7$, $mol/(dm^3 \cdot s)$	e	$W \cdot 10^7$, $mol/(dm^3 \cdot s)$	e
α – oxycyclohexylhydroperoxide					
308	0.91	0.72	0.63	1.44	0.32
318	2.07	1.76	0.59	3.28	0.31
328	3.91	3.96	0.49	7.04	0.28
340	9.50	10.0	0.48	16.6	0.29
α,α′ – dioxydicyclohexylperoxide					
313	0.32	0.31	0.52	0.6	0.27
328	1.27	1.24	0.52	2.2	0.29
338	2.87	2.84	0.51	4.96	0.29

α-Oxycyclohexylperoxides Decomposition Activated by $Et_4N^+Br^-$

The $Et_4N^+Br^-$ addition into the reaction medium in concentration which is one order less ($6.00 \cdot 10^{-3}$M) than peroxides concentration ($[ROOH]_0 = 5.00 \cdot 10^{-2}$M, $[ROOR]_0 = 5.24 \cdot 10^{-2}$M) leads to significant increasing of reaction rate as compared with peroxides thermal decomposition (343 K). The salt concentration during the reaction proceeding and in the end of reaction remains constant. This fact shows the catalytic character of tetraethylammonium bromide action upon α-oxycyclohexylperoxides decomposition.

Investigation of R^1OOR^2 decomposition in the presence of $Et_4N^+Br^-$ in pseudofirst order conditions on peroxides has shown the dependence of reaction rate effective constants from the ammonium salt concentration had nonlinear character (Fig. 2).

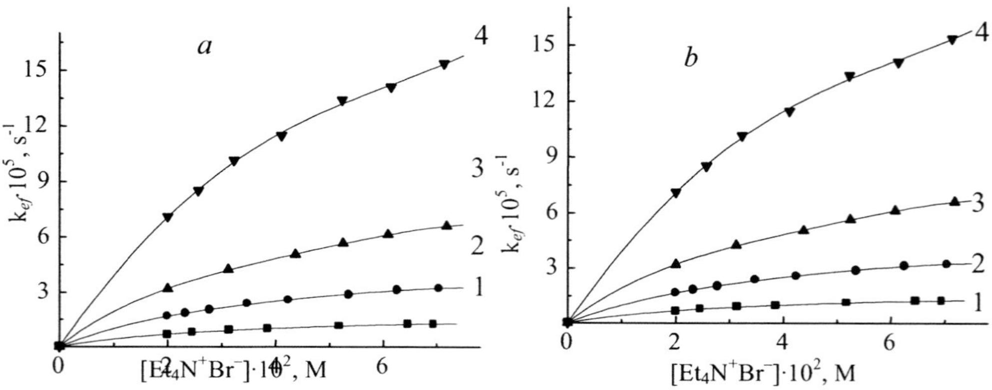

Figure 2. Dependence of effective rate constant of α – oxycyclohexylhydroperoxide (a) and α,α′ – dioxydicyclohexylperoxide (b) decomposition activated by tetraethylammonium bromide upon $Et_4N^+Br^-$ concentration ($[R^1OOR^2]_0 = 5 \cdot 10^{-3}$ M; T, K: 1 – 323, 2 – 333, 3 – 343, 4 – 353).

Investigation of the R^1OOR^2 decomposition within the following concentration interval $(0.6 - 5.0) \cdot 10^{-3}$ M in the presence of constant amount of $Et_4N^+Br^-$ ($3.21 \cdot 10^{-2}$ M) was carried out. The effective rate constant of the reaction of peroxides I and II with $Et_4N^+Br^-$ was found to be independent from the R^1OOR^2 concentration. This fact suggests that there are no any simultaneous paths of peroxide decomposition in the considered system.

The experimental facts point out onto the occurrence of complexation stage between the reactants. It is taken into account by the following kinetic scheme (equations 9 and 10):

$$R^1OOR^2 \xrightarrow{k_0} \text{products} \qquad (9)$$

$$R^1OOR^2 + Et_4N^+Br^- \underset{}{\overset{K_c}{\rightleftarrows}} \text{complex} \xrightarrow{k_d} \text{products} \qquad (10)$$

where k_0 – rate constant of R^1OOR^2 thermal decomposition s^{-1}; K_c – equilibrium constant of complexation between reactants, l/mol; k_d – rate constant of the complex decomposition reaction, s^{-1}.

The α – oxycyclohexylperoxides decomposition in the absence of ammonium salt have been investigated to estimate thermolysis contribution to the total reaction rate of R^1OOR^2 decomposition activated by $Et_4N^+Br^-$. The first order rate constants of R^1OOR^2 thermal decomposition in acetonitrile have been estimated:

α – oxycyclohexylhydroperoxide			
T, K	373	383	393
$k_0 \cdot 10^5$, s^{-1}	3.26 ± 0.09	9.6 ± 0.4	27.1 ± 0.6
α,α′ – dioxydicyclohexylperoxide			
$k_0 \cdot 10^5$, s^{-1}	1.45 ± 0.06	4.2 ± 0.2	13.6 ± 0.5

The temperature dependence of α – oxycyclohexylhydroperoxide and α,α′ – dioxydicyclohexylperoxide rate constants are described by the equations (11) and (12) correspondingly:

$$k_0 = (3.98 \pm 0.06)\cdot 10^{13}\exp((-129000 \pm 2000)/RT),\ c^{-1} \quad (11)$$

$$k_0 = (2.00 \pm 0.08)\cdot 10^{14}\exp((-136000 \pm 2000)/RT),\ c^{-1} \quad (12)$$

Since the rate of R^1OOR^2 thermal decomposition is significantly lower than the rate of their decomposition in the presence of bromide salt in conditions of reaction proceeding the effective rate constant is related with bromide salt concentration by the equation:

$$\frac{1}{k_{ef}} = \frac{1}{k_d K_C [Q^+ \cdot X^-]} + \frac{1}{k_d} \quad (13)$$

The dependence of k_{ef} on the $Et_4N^+Br^-$ concentration is linear in coordinates of equation (13) (Fig. 3). It is in the agreement with proposed kinetic scheme (10).

The values of equilibrium constants of complexation between peroxide compounds and tetraethylammonium bromide (K_c) and rate constants of complexed peroxide decomposition (k_d) are presented below:

α – oxycyclohexylhydroperoxide				
T, K	323	333	343	353
$k_d \cdot 10^5$, s^{-1}	5.1 ± 0.2	11.4 ± 0.4	25 ± 1	51 ± 2
K_c, mol/dm^3	44 ± 2	36 ± 2	29 ± 2	23 ± 1
α,α′ – dioxydicyclohexylperoxide				
$k_d \cdot 10^5$, s^{-1}	1.9 ± 0.1	5.2 ± 0.2	11.0 ± 0.6	28 ± 1
K_c, mol/dm^3	28 ± 3	23 ± 2	20 ± 2	17 ± 1

Values of enthalpy (ΔH_{com}) and entropy (ΔS_{com}) for α – oxycyclohexylhydroperoxide and α,α′ – dioxydicyclohexylperoxide complexation with tetraethylammonium bromide are -20 ± 1 kJ/mol and -30 ± 4 J/mol·K, -15 ± 3 kJ/mol and -19 ± 2 J/mol·K correspondingly, and value of free energy (ΔG_{com}, 323K) estimated on the base of presented data are -10.3 kJ/mol and -8.9 kJ/mol for peroxides I and II correspondingly. Temperature dependences of rate constants of associates of α – oxycyclohexylhydroperoxide and α,α′ – dioxydicyclohexylperoxide with $Et_4N^+Br^-$ decomposition are described by equations (14) and (15) correspondingly.

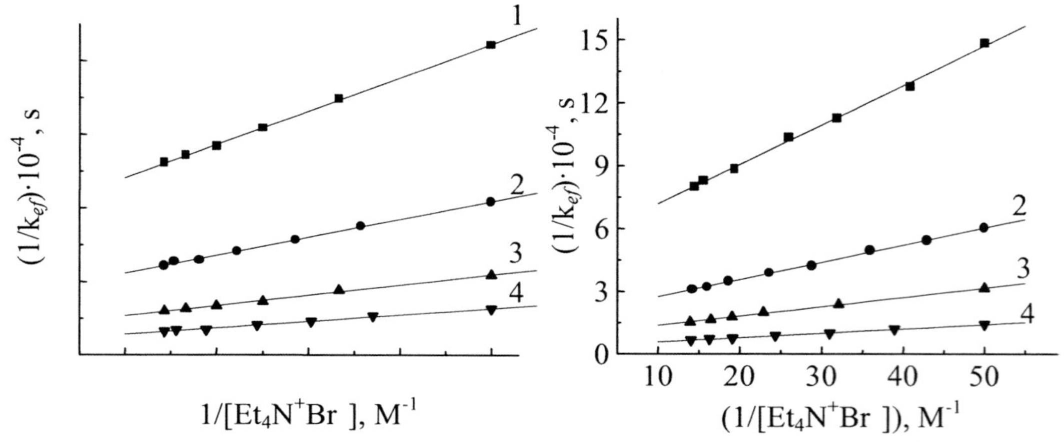

Figure 3. The dependence of effective reaction rate constant of α – oxycyclohexylhydroperoxide (a) and α,α′ – dioxydicyclohexylperoxide (b) decomposition activated by tetraethylammonium bromide in acetonitrile upon $Et_4N^+Br^-$ concentration in coordinates of equation (13) ($[R^1OOR^2]_0 = 5\cdot10^{-3}$ M; T, K: 1 – 323, 2 – 333, 3 – 343, 4 – 353).

$$k_d = (3.24 \pm 0.03)\cdot10^7\exp((-73000 \pm 1000)/RT),\ s^{-1} \quad (14)$$

$$k_d = (6.31 \pm 0.07)\cdot10^8\exp((-83000 \pm 2000)/RT),\ s^{-1} \quad (15)$$

Activation energy values $E_{a(k_d)}$, estimated from k_d (equations 14, 15) for both peroxides are some higher than corresponding values $E_{a(k)}$ determined under conditions of equal reagents initial concentrations in the system (equations 3, 5). In the proposed kinetic scheme (equation 10) the complexed peroxide decomposition with radicals products formation and catalyst regeneration is the rate determining stage. Temperature dependence of k_d allows to estimate the activation energy of complexed peroxide decomposition whereas temperature dependence of k yields total process energetics that undoubtedly takes into account heat of complexation (ΔH_{com}). But there is no additivity between activation enthalpy determined in corresponding conditions (Table 2) and complexation enthalpy:

$$E_{a(k)} \neq E_{a(k_d)} + \Delta H_{com} \quad (16)$$

And it can be seen taking into account the entropy contribution that compensative effect is observed in the case of free energy values (Table 2) that is in satisfactory agreement with the fact of complex formation in the system under consideration:

$$\Delta G^{\neq}_{k} = \Delta G^{\neq}_{k_d} + \Delta G_{com} \quad (17)$$

Table 2. Activation parameters of cyclohexanone peroxides decomposition activated by tetraethylammonium bromide

Conditions	$\Delta H^{\neq}$, kJ/mol	$\Delta S^{\neq}$, J/mol·K	$\Delta G^{\neq}$, kJ/mol (323K)
α – oxycyclohexylhydroperoxide			
k_d	70 ± 1	-111 ± 1	106 ± 1
k	64 ± 1	-102 ± 2	97 ± 1
α,α′ – dioxydicyclohexylperoxide			
k_d	81 ± 2	-86 ± 7	108 ± 5
k	72 ± 2	-90 ± 6	101 ± 4

It is obvious for both hydroperoxide and peroxide reaction activation barrier is lower as compared with thermolysis. Hydroperoxide is more active than peroxide. This fact can be explained perhaps by effect of peroxide compounds structure.

SUPRAMOLECULAR MODEL OF α OXYCYCLOHEXYLPEROXIDES DECOMPOSITION ACTIVATED BY TETRAETHYLAMMONIUM BROMIDE

The investigation of benzoyl peroxide decomposition in the presence of quaternary ammonium halides has shown that the nature of anion as well as cation of the onium salt influenced the peroxide substrate decomposition. Authors [19] have proposed the association model of diacylperoxides decomposition activation by onium salts (Fig. 4).

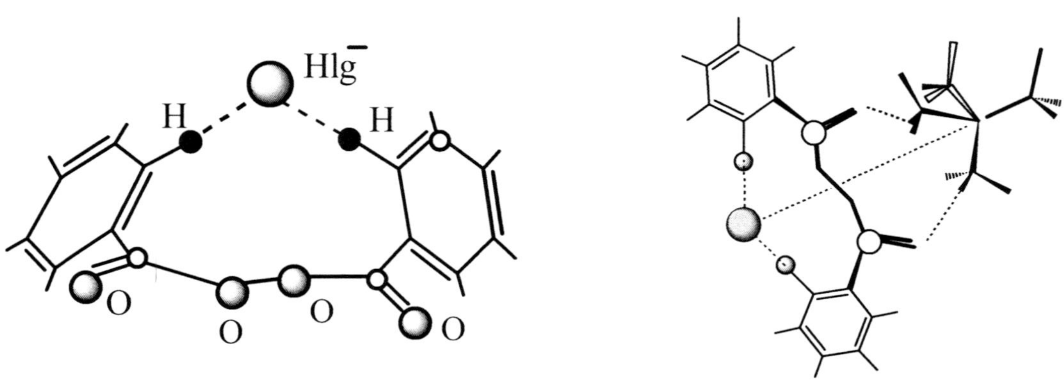

Figure 4. Associative models of benzoyl peroxide and onium halides.

We proceed from the conception that the decomposition of peroxide is possible if ion – molecular bonding of peroxide molecule by onium salt leads to molecule configuration changes and brings the peroxide structure to the structure of the decomposition reaction transition state. It leads to the decreasing of activation energy.

As the following equilibrium is typical for the $Et_4N^+Br^-$ in acetonitrile solution:

$$Et_4N^+Br^- \rightleftarrows Et_4N^+ + Br^- \quad (18)$$

the quantum chemical modeling of cyclohexanone peroxide compounds associates (19-21) with $Et_4N^+Br^-$ ions was carried out using AM1 method approximation realized in the program package MOPAC – 2000 [20].

$R^1OOR^2 ... Br^-$	(19)
$Et_4N^+ ... R^1OOR^2$	(20)
$Et_4N^+ ... R^1OOR^2 ... Br^-$	(21)

α,α′ – dioxydicyclohexylperoxide has two oxy-groups with which bromide ions potentially can associate on hydrogen bonding type. The peroxide molecule has the C_2 – symmetry. The dipole moment vector is directed along symmetry axis, its negative pole is located in the center of peroxide bond and positive one in such case could determine the associative interaction coordinate of peroxide with bromide – anion (Fig. 5). That is why the peroxide symmetry axis is chosen as association coordinate.

Investigation of benzoyl peroxide decomposition activated by onium halides do not let exclude the cation effect on the kinetics of peroxide decomposition (Fig. 6). Two possible ways of association between tetraethylammonium – cation and peroxide along symmetry axis have been considered (Fig.5).

The investigations of benzoyl peroxide decomposition kinetics with the tetraethylammonium perchlorate assistance have shown the salt to be kinetically inert [8]. Taking into account that this salt is completly dissosiated in acetonitrile we can conclude that the cation does not participate in the reaction. Thus the only cation action could not lead to peroxide activation.

Assosiates models which can be formed uder combined action of the onium salt ions are presented on (Fig.7). The associative interactions enthalpies of corresponding tetraethylammonium bromide ions with peroxide are –74.3, -31.2, -17.9, -41.2 kJ/mol correspondingly (Fig. 5). In the case of two ions and peroxide the association enthalpies are -59.6, -20.2 kJ/mol (Fig.7). We assume that the such type associate is most likely formed by exchange mechanism according with scheme (22):

$$Et_4N^+...Solv...Br^- + ROOR \longrightarrow Et_4N^+...ROOR...Br^- + Solv \quad (22)$$

The corresponding reaction enthalpy values are ΔH = -53.7 kJ/mol and -20.2 kJ/mol correspondingly for the of α – oxycyclohexylhydroperoxide and α,α′ – dioxydicyclohexylperoxide.

So, any associate (I-V) can be formed. The calculations show that -O-O- bond order is decreasing, for example, in anion – peroxide associate (model II) whereas in associate I ΔH_{as} value increases.

From our point of view -O-O- activation is observed in associate II, although the associate I formation is more favorable. Peroxide bond activation is observed only in that case when association of anion with peroxide is accompanied by significant changes of COOC torsion angle: it changes from equilibrium (119.6^0) to antiperiplanar (151.5^0). Associate I formation is accompanied by conformation changes of torsion angles containing OH-groups, which participates in the ion – molecular hydrogen bond formation. At the same time the COOC molecular fragment changes slightly (from 119.6^0 to 113.4^0) and peroxide bond become even stronger (Table 3). On the base of bond order changes preference can be given to the associates II, IV, VI because in these cases -O-O- bond order decreasing is observed. There is no activation of peroxide molecule in the case of peroxide – onium cation associate formation. The peroxide molecule configuration is not changed though there are no significant changes in the electron structure of the peroxide.

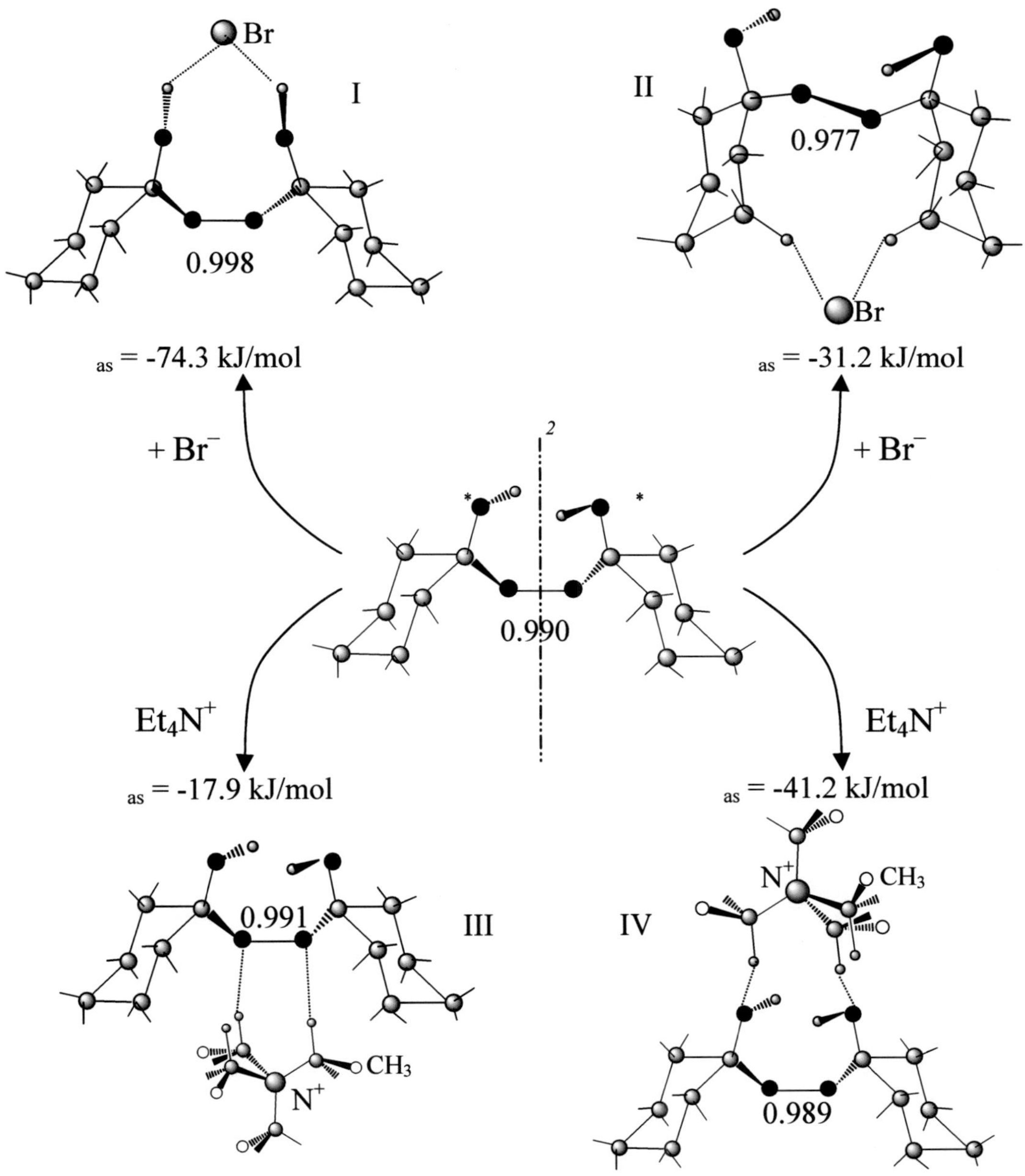

Figure 5. Stereochemical models of α,α′ – dioxydicyclohexylperoxide and their possible associates. (The -O-O- bond order is presented).

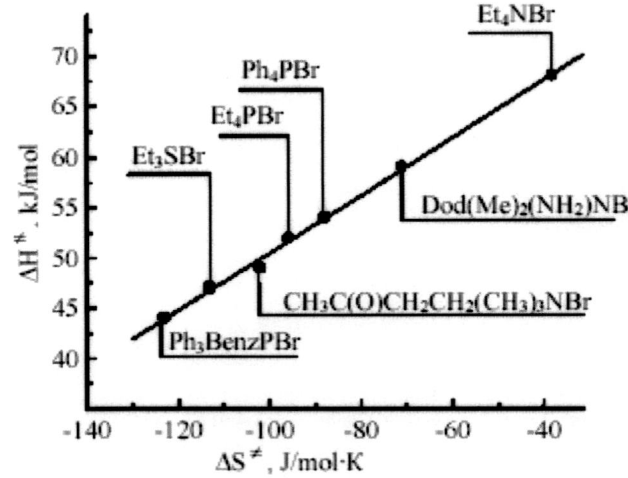

Figure 6. The effect of cation on the activation parameters of benzoyl peroxide decomposition reaction in acetonitrile.

Table 3. α,α′ – dioxydicyclohexylperoxide and its associates characteristics

Object	Associate	ΔH^0_f, kJ/mol	∠COOC, degrees	[1]ΔH^0_f, kJ/mol	[2]E_{-O-O-}, eV
R(OH)COOC(OH)R	-	-737.8	119.6	-737.8	-11.64
ROOR ... Br^-	I	-897.5	113.4	-692.4	-11.94
	II	-854.4	151.5	-716.3	-11.14
Et_4N^+ ... ROOR	III	-202.4	111.1	-731.4	-11.75
	IV	-225.7	121.1	-735.3	-11.64
Et_4N^+ ... ROOR... Br^-	V	-556.7	94.8	-662.0	-11.87
	VI	-517.3	181.8	-691.9	-11.04

[1] ΔH^0_f – standard formation enthalpy of peroxide.
[2] E_{-O-O-} - energy of two-atomic interactions in Malliken approximation.

Only under combined action of anion and cation the most essential changes of the peroxide fragment (C-O-O-C) conformation in the ROOR molecule are observed. It is the dominant factor of peroxide bond activation. Considering models V and VI it is impossible to prefer finally only one structure. But comparative analysis of all considered models allows conclude about selective association of peroxide with tetraethylammonium bromide ions; i.e. catalyst attack should be stereospecific.

Analogous approach was used to develop the stereochemical models of α-oxycyclohexylhydroperoxide and tetraethylammonium bromide associates. The hydroperxide structure changes are also observed in the case of such associate formation (Fig. 8).

Thus, ion – molecular associate formation between peroxide and onium salt should follow reactants stereo-specific orientation.

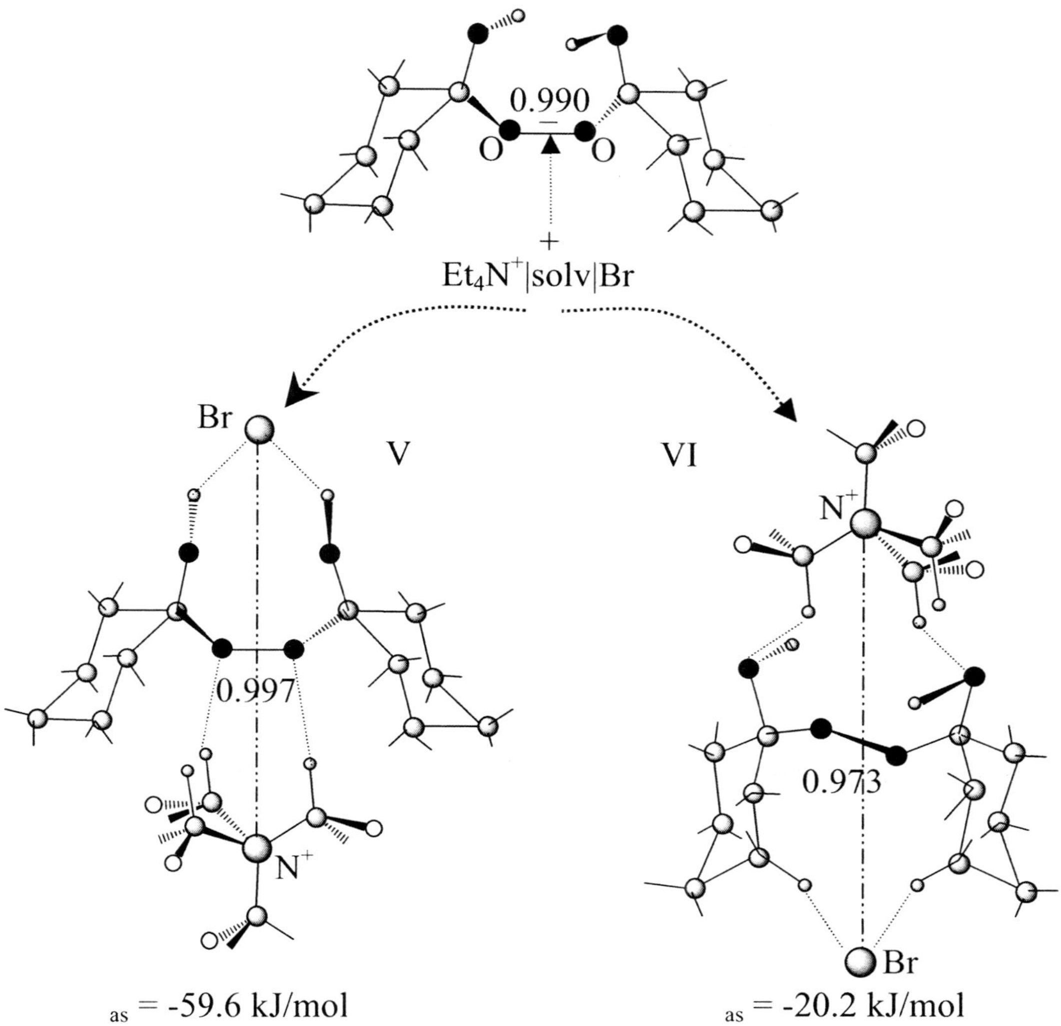

Figure 7. Associative model of α,α′ – dioxydicyclohexylperoxide with tetraethylammonium bromide (-O-O- bond order is presented).

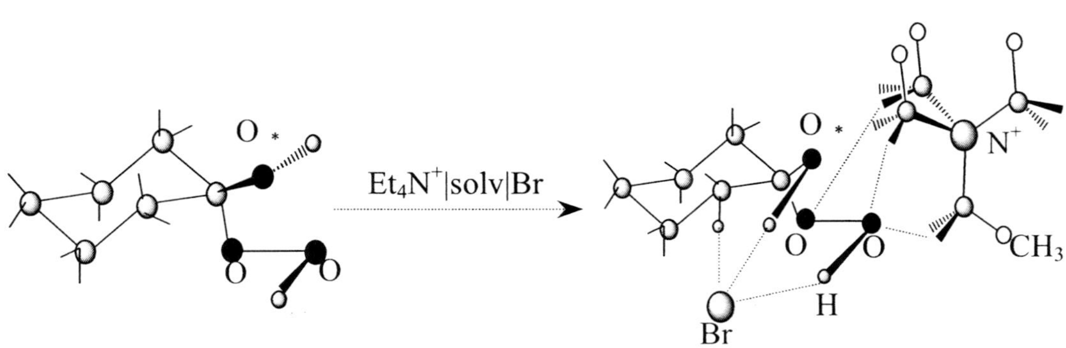

Figure 8. Associative model of α – oxycyclohexylhydroperoxide with tetraethylammonium bromide.

On the base of presented data the following scheme of cyclohexanone peroxides activation and decomposition in the presence of tetraethylammonium bromide can be proposed:

+ Solv

$$Q^+ \,||\, X^- + ROOR' \rightleftarrows Q^+...ROOR'...X^- \longrightarrow [Q^+...ROOR'...X^-]^{\neq} \longrightarrow RO^{\bullet} + {}^{\bullet}RO + Q^+ \bullet X^-$$

In the case of complex formation peroxide geometry and electron structure changes bring it to the configuration close to that of transition state. Complex decomposition proceeds through oxyradical formation and ammonium salt regeneration. Assumed mechanism of cyclohexanone peroxides decomposition in the presence of tetraethylammonium bromide can be regarded as one of the supramolecular reactions [21].

References

[1] V.L. Antonovskij. *Progress in the chemistry of organic peroxides*. Moscow, TsNIITEneftekhim, 1992.

[2] W. Liu, X. Liu, D. Knaebel, L.Li.Y.Luck. *Antimicrob. Agents Chemother*. 42, № 4. 911 (1998).

[3] T. Tsubota, S. Ida, O. Hirabayashi, S. Nagaoka, M. Nagata, Y. Matsumoto. *Phys. Chem. Chem. Phys*. 4. 3881 (2002).

[4] N.A. Turovskyj, S.U. Tselinskyj, I.A. Opeida, R.V.Kucher. *Dokl. Akad. Nauk USSR*, №3, 131 (1991).

[5] N.A. Turovskyj, S.U. Tselinskyj, I.A. Opeida, U.E. Shapiro. *Teor. Eksp. Khim*. 28, №4. 324 (1992).

[6] N.A. Turovskyj, S.U. Tselinskyj, U.E. Shapiro, A.R. Kaluskyj. *Teor. Eksp. Khim*., 28, №4. 320 (1992).

[7] N.A. Turovskyj, S.U. Tselinskyj. *Ukr. Khim. Zh*., 60, №1. 16 (1994).

[8] N.A. Turovskyj, S.U. Tselinskyj, I.A. Opeida. *Teor. Eksp. Khim*., 31, №1. 52 (1995).

[9] S.U. Tselinskyj, N.A. Turovskyj, I.A. Opeida, E.L. Baranovskyj. *Teor. Eksp. Khim.,* 32, №2. 88 (1996).

[10] I.A. Opeida, N.A. Turovskyj, N.V. Maksuta. *Zh. Org. Khim*., 36, Iss. 11. 1695 (2000).

[11] E.M. Gavryliv, N.A. Turovskyj, R.V. Kucher. *Dokl. Akad. Nauk* USSR 11. 32 (1990).

[12] N.A. Turovskyj, S.U. Tselinskyj, I.A. Opeida, A.N. Nikolayevskyj, D.L. Hmelnitskaya, E.L. Baranovskyj, M.U. Zubritskyj. *Ukr. Khim. Zh*., 61, №5. 67 (1995).

[13] I.A. Opeida, N.A. Turovskyj, N.V. Maksuta, A.I. Pomeschenko. *Kinetika i Kataliz*, 42, №5. 1 (2001).

[14] I.A. Opeida, N.M. Zalevskaya, E.N. Turovskaya, U.I. Sobka, *Petroleum Chemistry*, 42, №6. 460 (2002).

[15] I.A. Opeida, N.M. Zalevskaya, E.N. Turovskaya. *Teor. Eksp. Khim.,* 39, №4. 236 (2003).

[16] I.A. Opeida, N.M. Zalevskaya, E.N. Turovskaya. *Petroleum Chemistry*. 44, №5. 358 (2004).

[17] Vaisberger, E. Proskauer, J. Ruddik, E. Tups. *Organic Solvents.* [Russian Translation]. Izd. Inostr. Lit., Moscow, 1958. A.

[18] A.J. Gordon, R.A. Ford. *The Chemist's Companion.* [Russian Translation]- Mir, Moscow, 1976.

[19] N.A. Turovskyj, I.A. Opeida, A.N. Nikolayevskyj, V.L. Antonovskyj. *Tez. Dokl. Tr. Vserossijskoj konferentsii "Molekul'arnoe modelirovanie"*, Moscow. 125 (2003).

[20] J.J.P. Stewart. *MOPAC program package: Manuel.* 2000.

[21] J.-M. Lehn. *Supramolecular Chemistry: Concepts and Perspectives.* [Russian Translation]. Novosibirsk: Nauka. Sib. Predpri'atie RAN, 1998.

In: Order and Disorder in Polymer Reactivity
Editors: G. E. Zaikov and B. A. Howell, pp. 53-75
ISBN 1-60021-145-3

Chapter 6

INTERACTION OF POLYMERS WITH THE NITROGEN OXIDES IN POLLUTED ATMOSPHERES

G. B. Pariiskii, I. S. Gaponova and E. Y. Davydov
Institute of Biochemical Physics, Russian Academy of Sciences
4 Kosygin str., Moscow 119991, Russia

1. INTRODUCTION

In this paper, the mechanisms of the reactions of nitrogen oxides with solid polymers are considered. Active participants in reactions with nitrogen oxides are double bonds, the amide groups of macromolecules, alkyl, alkoxy and peroxy radicals, as well as hydroperoxides. The structure of the reaction front during nitration of rubbers has been studied using the electron spin resonance (ESR) imaging technique. The reactions with nitrogen oxides provide a simple way of preparing spin-labelled polymers. The structural-physical effects on the kinetics and mechanism of reactions of nitrogen dioxide have been demonstrated by the example of filled polyvinylpyrrolidone (PVP).

Thermal and photochemical oxidation of polymers have been the subject of detailed and prolonged investigations, because these processes are of major importance for the stabilisation of polymeric materials. However, since the 1960s, the influence of aggressive gases in polluted atmospheres on polymer stability has attracted considerable attention [1]. Among such pollutants in the atmosphere, sulfur dioxide, ozone and the nitrogen oxides stand out as the most deleterious. However, the pursuance of this research has run into a number of problems. The interaction of pollutants with polymers involves the penetration of gases into solids and thus results in a complex kinetic description of the process. Also, as a rule, these reactions are long term for the concentrations of pollutants found in the environment. Consequently, other aging processes occur in the actual conditions of use and storage of polymer materials.

To establish the effect of a given aggressive gas on a particular polymer, the reaction is generally studied at pollutant concentrations that are much higher than those actually existing in polluted atmospheres. The results obtained by this means are then linearly extrapolated to the concentrations of reactants found in the atmosphere. This expedient is, *a priori,* ambiguous in view of the fact that the role of the individual stages of a uniform aging process is changed in conditions of accelerated testing.

The problem of non-equivalent kinetics is inherent to polymer reactions in solids [2]. In this case particles existing in different surroundings react with different rate constants. As a result, the most active particles will be removed from the reaction, and the overall rate constant will decrease with time. On the other hand, relaxation processes in polymers restore the initial distribution of particles and so their reactivity. Thus the kinetics will depend on the relation between the rate of the chemical reaction and the rate of the relaxation processes [3], This fact also makes it necessary to reconsider critically the validity of extending the results of accelerated tests for polymer ageing.

This chapter is devoted to a consideration of the results obtained in studies of the interactions of nitrogen oxides with polymers. There are eight nitrogen oxides, but only NO, NO_2 and N-,O_4 are actually important as pollutants. Nitric oxide (NO) exists as a free radical, but it is reasonably stable in reactions with organic compounds. The paramagnetic nitrogen dioxide (NO_2) is more active compared with NO. This gas is universally present in equilibrium with its dimer molecule:

$$2NO_2 \rightleftharpoons N_2O_4$$

with K_p = 0.141 arm at 298 K [4]. Nitrogen dioxide absorbs light in the near-UV and visible spectral range. Excited molecules are generated by light with $\lambda > 400$ nm. The dissociation of NO_2 into an oxygen atom and NO by light with $\lambda < 365$ nm takes place with a quantum yield near to unity [5].

2. Interaction of Nitrogen Dioxide with Polymers

Detailed investigations of the reactions of NO_7 with various polymers have been carried out by Jellinek and co-workers [1, 6]. The degradation of polymer films has been studied at different pressures of NO_2, in mixtures of NO_2 with air, under the combined action of light ($\lambda > 280$ nm), O_2 and NO_2. Based on the data obtained, Jellinek classified all polymers into three groups:

1. vinyl polymers - polyethylene (PE), polypropylene (PP), polystyrene (PS), polymethyl methacrylate (PMMA), polyacrylonitrile (PAN), polyvinyl chloride (PVC) and polyvinyl fluoride (PVF);
2. polymers with non-saturation - primarily rubbers;
3. polyamides, polyurethanes and polyamidoimides.

The presentation of the results in this section will be carried out according to this classification.

2.1. Vinyl Polymers: PE, PP, PS, PMMA, PAN, PVC and PVF

The linear extrapolation of the results of accelerated tests to NO_2 concentrations likely to be found in the atmosphere (1-5 ppm) predicts that polymer properties will be essentially constant for a long time.

The first investigations of the interaction of NO_2 with PE and PP were performed by Ogihara and co-workers [7, 8] at 298-383 K and NO_2 pressure of 20 kPa. It was established that NO, reacts at room temperature with the >C=C< double bonds originally contained in PE with the formation of dinitro compounds and nitronitrites by the following reactions:

$$>C=C< + NO_2 \rightarrow >C^{\bullet}-C(NO_2)< \quad (= R_1^{\bullet}) \tag{1}$$

$$R_1^{\bullet} + NO_2 \rightarrow >C(NO_2)-C(NO_2)< \tag{2}$$

$$R_1^{\bullet} + ONO \rightarrow >C(ONO)-C(NO_2)< \tag{3}$$

Hydrogen atom abstraction does not take place at room temperature. The nitro, nitrite, nitrate, carbonyl and hydroxyl groups are formed at T > 373 K. The following mechanism was postulated:

$$RH + NO_2 \rightarrow R^{\bullet} + HNO_2 \tag{4}$$

$$R^{\bullet} + NO_2 \rightarrow RNO_2 \tag{5}$$

$$R^{\bullet} + ONO \rightarrow RONO \tag{6}$$

$$RONO \rightarrow RO^{\bullet} + NO \tag{7}$$

The reactions of $RO^{\bullet}$ radicals lead to the formation of macromolecular nitrates, alcohols and carbonyl compounds. The activation energy of the NO_2 addition to the double bonds of PE is 8-16 kj/mol. The activation energies of H atom abstraction are 56-68 kj/mol in PE and 60 kj/mol in PP.

PE, PP, PAN and PMMA change their characteristics slightly at high concentrations of NO_2 (1.3-13 kPa) even under the joint action of pollutant, O_2 and UV light [6]. Nitrogen dioxide is capable of abstracting tertiary hydrogen atoms in PS with a low rate (P = 20-80 kPa), with the formation of nitro and nitrite side groups [reactions (5) and (6)]. This process is accompanied by main-chain scission [9, 10]. The combined action of 0.3 kPa NO_2 and light (λ > 280 nm) on PS does not lead to main-chain decomposition in the early stage (10 h), after which the degradation process is developed with a constant rate. PVC and PVF show a minor loss of chlorine and fluorine atoms on exposure to NO_2 [1, 6].

An attempt to investigate quantitatively the ageing of PS and poly-tert-butyl methacrylate (P-*t*-BuMA) has been taken by Huber [11]. The research was performed in a flow system of air containing 60-900 ppm of NO_2 and/or 60-900 ppm SO_2 at 300 K under the simultaneous action of light with $\lambda > 290$ nm. The degradation of P-*t*-BuMA films was expressed in terms of the quantity of ruptures per 10,000 monomer units, α. The kinetic dependence is represented by the equation:

$$\alpha = (P/Q)[\exp(Qt) - 1] \tag{8}$$

where P and Q are constants. This equation describes an autoaccelerated process. As $Q \to 0$, so $\alpha \to Pt$, that is, the degradation proceeds with a constant rate. The P and Q values decrease as the film thickness increases, and yet the P value diminishes more strongly than Q. Therefore, the accelerated character of the degradation appears more clearly for thin films. PS degradation in the same conditions proceeds much more slowly and has a more pronounced autoacceleration (Table 1).

Table 1. The P and Q values for P-*t*-BuMA and PS film degradation under the action of 100 ppm NO_2 and light in air

Polymer type	Film thickness (mg/cm^2)	$P \times 10^4 (h^{-1})$	$Q \times 10^4 (h^{-1})$
P-*t*-BuMA	1.4	0.071	0.026
P-*t*-BuMA	2.6	0.050	-
P-*t*-BuMA	2.8	0.041	0.017
PS	1.4	0.034	0.036

The autoaccelerated character of P-*t*-BuMA degradation was linked to the ester group decomposition, with isobutylene formation, which gives free radicals in the reaction with NO_2 and thus promotes the degradation process.

The IR spectrum of PS shows peaks corresponding to carbonyl {1686 cm^{-1}) and hydroxyl (3400 cm^{-1}) groups after exposure to a mixture of NO_2 (100 ppm) and air. No bands connected with the insertion of NO_2 into the P-*t*-BuMA and PS macromolecules were observed. It is believed that the following sequence of reactions occurs in PS [11]:

$$RH + NO_2 \rightarrow R^{\bullet} + HNO_2 \tag{9}$$

$$R^{\bullet} + O_2 \rightarrow RO_2^{\bullet} \tag{10}$$

$$RO_2^{\bullet} + RH \rightarrow ROOH + R^{\bullet} \tag{11}$$

$$R^{\bullet} + NO_2 \rightarrow RNO_2 \quad (12)$$

$$R^{\bullet} + NO_2 \rightarrow RONO \quad (13)$$

$$RONO \rightarrow RO^{\bullet} + NO \quad (14)$$

$$ROOH + NO \rightarrow RO^{\bullet} + {}^{\bullet}OH + NO \quad (15)$$

$$ROOH + h\nu \rightarrow RO^{\bullet} + {}^{\bullet}OH \quad (16)$$

$$RO^{\bullet} \rightarrow \text{degradation} + R^{\bullet} \quad (17)$$

Hydroperoxide decomposition under the action of NO and light gives rise to accelerated PS degradation.

2.2. Non-Saturated Polymers

Among these are primarily rubbers. These polymers are far more sensitive to NO_2 action that the polyolefins. Appreciable degradation of macromolecules as well as moderate crosslinking were observed for rubbers.

Comprehensive kinetic investigations of butyl rubber (a copolymer of isobutylene with 1.75% isoprene) in an NO_2 atmosphere (NO_2 pressure 1.33-133 kPa), in a mixture of NO_2 and air, and under the combined action of NO_2, O_2 and UV light (λ > 280 nm) have been performed by Jellinek and co-workers [12, 13]. According to the proposed mechanism, the total number of chain ruptures is made up of three parts: (1) ruptures that are due to the NO_2 interaction only, (2) ruptures that result only from the action of O_2, and (3) ruptures that are caused by the combined action of NO_2 and O_2.

The kinetic dependence of the degree of degradation, a= (I/DP,- 1/DP$_O$), is described by the following equation:

$$\alpha = k'_{ef} t^2 + k''_{ef} [NO_2][1 - \exp(-k_3 t)] \quad (18)$$

where DP$_0$ and DP, are the number-average degrees of polymerisation in the original and degraded macromolecule (at time t). The first term in equation (18) is connected with ruptures of macromolecules due to photolysis of the reaction products (hydroperoxides, nitro and nitrite groups). The second term describes the degradation for the (NO_2 + O_2) system in the absence of light.

It should be noted that the assumed mechanism [12, 13] is very complex, involving a wealth of elementary reactions, the rate constants of which are unknown in the solid phase. It is well known that the reaction products can be more active relative to the nitrogen oxides than the original polymer. In connection with this, the application of various physical-chemical techniques is extremely important to investigate the degradation process. The development of methods to study the movement of the reaction front across the polymer sample is also required. The use of the ESR technique permits one to draw additional conclusions on the mechanisms of the interaction of polymers with nitrogen oxides from the structure of the resulting free radicals and the kinetics of their formation.

The interaction of polyisoprene (PI) with NO_2 gives rise to di-*tert*-alkylnitroxyl radicals [14]. The ESR spectra of these radicals show a characteristic anisotropic triplet signal with a width of $2A^N_{||} = 6.2$ mT and $g_{||} = 2.0028 \pm 0.0005$ in the solid polymer, and a triplet with $a^N = 1.53 \pm 0.03$ mT and $g = 2.0057 + 0.0005$ in dilute solutions. These macroradicals are stable in the absence of NO_2 during storage for many months in both inert atmosphere and air. The proposed scheme to explain the formation of these radicals involves three main stages: (1) generation of TV-containing alkyl radicals, (2) synthesis of tertiary macromolecular nitroso compounds, and (3) spin-trapping of the tertiary alkyl or allyl radicals:

$$\text{\textasciitilde}CH_2\text{-}C(CH_3)\text{=}CH\text{-}CH_2\text{\textasciitilde} + NO_2 \rightleftharpoons \text{\textasciitilde}C^{\bullet}(CH_3)\text{-}CH(ONO)\text{-}CH_2\text{\textasciitilde} \quad (19)$$

$$\text{\textasciitilde}C^{\bullet}(CH_3)\text{-}CH(ONO)\text{-}CH_2\text{\textasciitilde} + RH \rightleftharpoons \text{\textasciitilde}C(CH_3)(NO)\text{-}CH(OH)\text{-}CH_2\text{\textasciitilde} \quad (20)$$

$$\text{\textasciitilde}C(CH_3)(NO)\text{-}CH(OH)\text{-}CH_2\text{\textasciitilde} + {}^{\bullet}R_{tert} \rightarrow R_{tert}\text{-}N(O^{\bullet})\text{-}R_{tert} \quad (21)$$

The reactions of NO_2 with double bonds provide a very simple and rapid method for the synthesis of spin-labelled macromolecules of rubbers. The temperature variation of the rotational mobility of macromolecules in block PI has been studied using spin-labelled samples [14]. The temperature dependence of the rotational correlation time τ is described by $\tau_c = \tau_0 \exp(J/RT)$. The τ_c values within the fast motion region ($\tau_c < 10^{-9}$ s) are well described by the parameters $E = 34.7$ kj/mol and $\log \tau_0 = -14.2$.

The spatial distribution of these macromolecular nitroxyl radicals allows the estimation of the spatial distribution of the nitration reaction in bulk PI. The possibilities of the ESR imaging technique to determine the form of the reaction front of PI nitration has been considered [15]. The ESR imaging spectra were registered in an inhomogeneous magnetic field on cylindrical samples of 0.4 cm diameter and 1 cm height at NO_2 and O_2 concentrations of 1×10^{-4} to 2×10^{-3} mol/1 and 2×10^{-3} to 1.4×10^{-2} mol/1, respectively. The spatial distributions of $R_2NO^{\bullet}$ radicals at various reaction times are shown in Figure 1. The width of the distribution varies over 20-30% for 740 h. The maximum concentration of nitroxyl radicals is observed in the superficial layer, and it progressively decreases towards the centre. The width of this layer is ~1 mm, and radicals are unavailable in the sample centre. The nitroxyl radical yield with respect to absorbed NO-, molecules is 0.01. The shape and variation of the distribution in the presence of O_2 are the same as in pure NO_7, but the reaction front is narrower. The rate of $R_2NO^{\bullet}$ formation in the presence of O_2 is much lower than in pure NO_2 at the cost of a decay of alkyl radicals in the reactions with O_2: $W(NO_2)/W(NO_2{+}O_2) = 10^2$. The distribution at a fixed distance from the surface is likely determined by macrodefects in the sample volume, namely, the availability of cracks and porosity. The front form is determined by the 'membranous' regime of the nitration process rather than by structural changes.

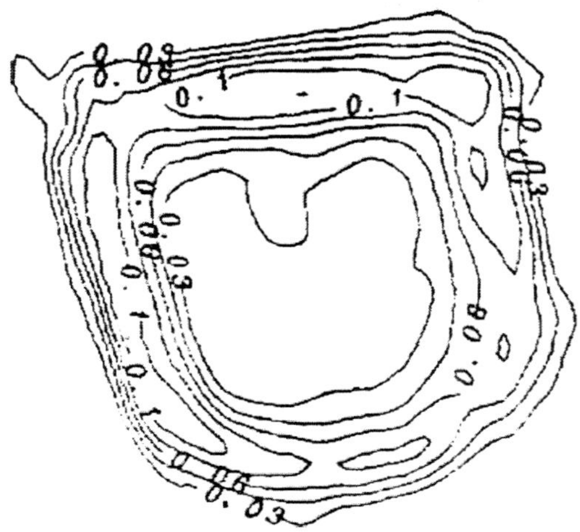

Figure 1. ESR-imagination of nitroxyl radicals distribution in cylindrical sample (l = 10 mm, d = 4 mm) of PI in course of interaction with nitrogen dioxide ([NO_2]) = 8.8 × 10^{-4} mole/1; 30 min; 20 °C). The contour lines correspond to vertical sections with equal [$R_2NO^{\bullet}$]. The concentrations are given in arbitrary units ([$R_2NO^{\bullet}$]$_{max}$ = 0.125 au).

PMMA, which in itself is stable on exposure to NO_2, enters into reactions after previous irradiation by UV light at 293 K [16]. The photolysis of PMMA induces the formation of double bonds as a result of ester group decomposition. The ESR spectrum observed after exposure of samples to NO_2 is shown in Figure 2. The spectrum represents the superposition of the signals of two nitroxyl radicals at low frequencies of rotational mobility (10^{-9} s < τ_c < 10^{-7} s):

- Dialkylnitroxyl radicals

$$\sim C(CH_3)(COOCH_3)\text{–}N(O^{\bullet})\text{–}C(CH_3)(CHO)\text{–}CH_2\sim$$

give an anisotropic triplet signal with hyperfine interaction (HFI) constant $A^N_{||}$ = 3.2 ± 0.1 mT and $g_{||}$ = 2.0026 ± 0.0005;

- Acylalkylnitroxyl radicals

$$\sim C(CH_3)(COOCH_3)\text{–}CH(OH)\text{–}C(CH_3)[N(O^{\bullet})COOCH_3]\text{–}CH_2\sim$$

give a triplet signal that is characterised by $A^N_{||}$ = 2.1 ± 0.1 mT and $g_{||}$ = 2.0027 + 0.0005.

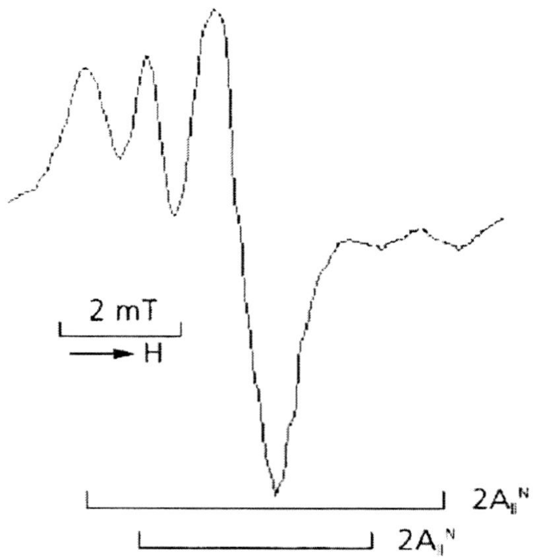

Figure 2. ESR spectrum of nitroxyl radicals generated by NO, in PMMA pre-irradiated by UV light at 298 K.

The free-radical process of NO_2 interaction with PMMA containing double bonds is represented by the scheme opposite.

The formation of nitroxy] radicals testifies to the fact that main-chain decomposition by reaction (24) and side-chain ester group cleavage by reaction (26) take place in the polymer. Thus, the interaction of NO-, with double bonds is able to initiate free-radical reactions of polymer degradation when hydrogen atom abstraction reactions from C-H bonds are inefficient.

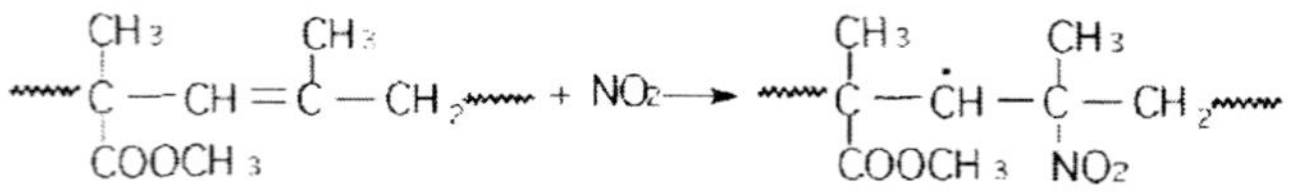

(22)

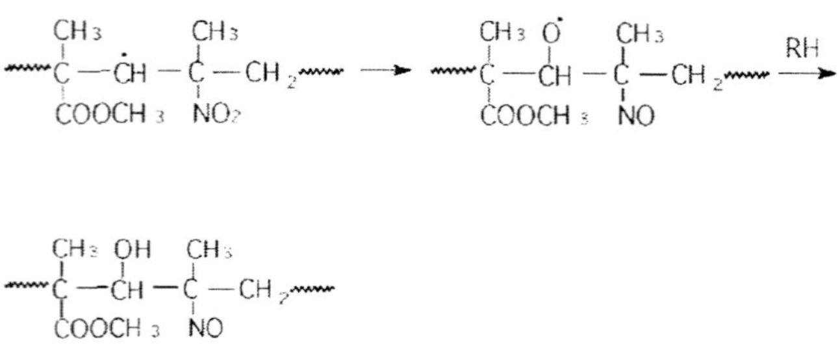

(23)

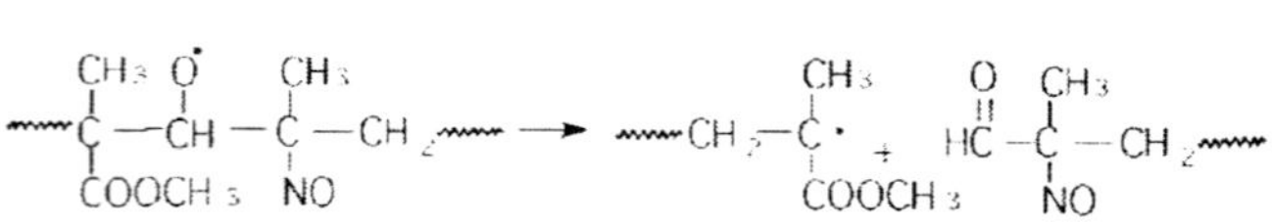

(24)

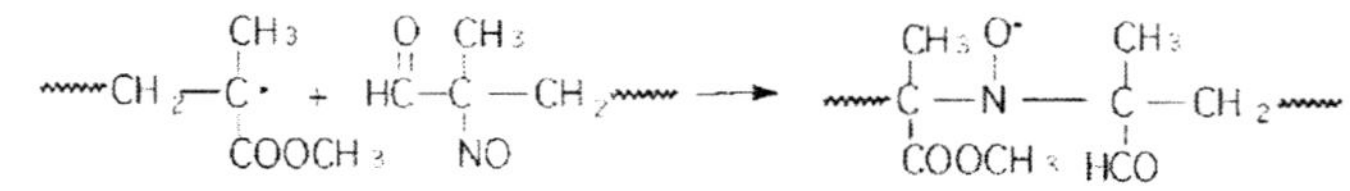

(25)

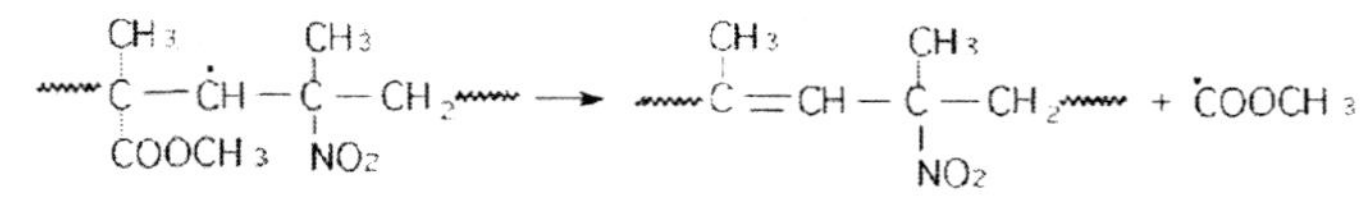

(26)

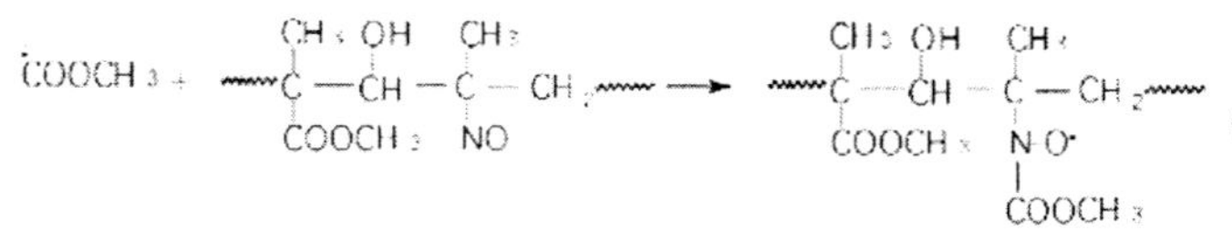

(27)

2.3. Polyamides, Poiyurethanes, Polyamidoimides

Polymers with amide and urethane groups in the macromolecules represent a special class of materials that are sensitive to NO_2. The action of NO_2 at pressures of 0.5-2 mm Hg on polyamide-6,6 films with various morphologies has been studied by Jellinek and co-workers [17, 18]. It was shown that a degradation process takes place. The degradation of polyamide is a diffusion-controlled reaction and depends on the degree of crystallinity and the sizes of the crystallites. The process is inhibited by small quantities of benzaldehyde or benzoic acid. Increase of the degradation rate was observed during the combined action of NO_2, air and UV light. The assumed mechanism of the process as follows:

$$\text{-CO-NH-} + NO_2 \rightarrow \text{-CO-N}^{\bullet}\text{-} + HNO_2 \qquad (28)$$

$$\text{-CO-N}^{\bullet}\text{-} + NO_2 \rightarrow \text{-CO-N}(NO_2)\text{-} \qquad (29)$$

$$\text{-CO-N}^{\bullet}\text{-CH}_2\text{-} \rightleftharpoons [\text{-CO-N=CH}_2 + {}^{\bullet}\text{CH}_2\text{-}] \rightarrow \text{chain rupture} \qquad (30)$$

There is reason to believe that only a small quantity of amide groups, not linked by the hydrogen bonds, enter into the reaction. These groups can be interlocked by benzoic acid with the formation of the following structure:

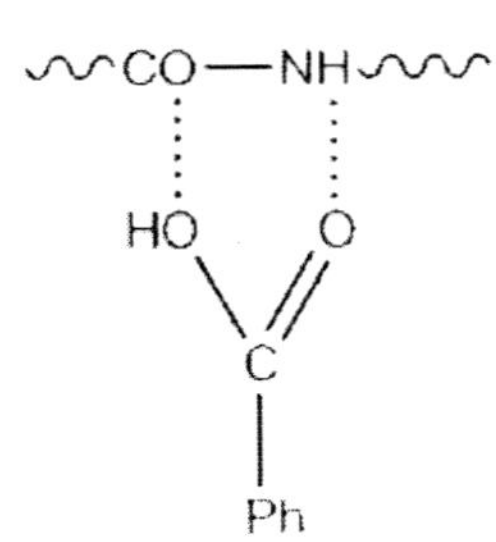

Research into the effect of NO_2 on polyamide textiles has been described [19], The exposure of samples in an NO_2 atmosphere of low concentration at room temperature for 100 h does not lead to a decrease in the whiteness and tensile strength. However, these characteristics are decreased at higher temperatures. The availability of nitrogen oxides in the air under the action of UV light results in the additional degradation of textiles.

The conversion of N–H bonds by nitrogen dioxide is also inherent to polycaproamide (PCA). The UV spectra of PCA films display features of absorption at 390-435 nm during exposure to NO_2 at concentrations of 10^{-4} to 10^{-3} mol/l [20]. The absorption bands were assigned to nitrosamide groups resulting from N–H group conversion. This conclusion is confirmed by IR spectroscopy. The intensity of the hand with ν = 3293 cm^{-1}, which is associated with stretching vibrations of the hydrogen-linked N-H groups, decreases sharply. The intensities of the amide I (ν = 1642 cm^{-1}) and amide II (ν = 1563 cm^{-1}) bands, which are characteristic of PCA, also decrease. Instead of these bands, absorptions at ν = 1730 cm^{-1}, which corresponds to the absorption of C=O groups, and at ν = 1504 and 1387 cm^{-1}, which correspond to stretching vibrations of N=O groups of nitrosamides, appear in PCA. Thus, nitrosation through the amide group is the main process of PCA transformation in an NO_2 atmosphere, which leads to disintegration of the system of hydrogen bonds. Taking into account the equilibrium:

$$NO^+NO_3^- \rightleftharpoons N_2O_4 \rightleftharpoons 2NO_2 \qquad (31)$$

the formation of nitrosamides can be represented as follows:

$$\text{~CONHCH}_2\text{~} + N_2O_4 \rightarrow \text{~CON(NO)CH}_2\text{~} + HNO_3 \qquad (32)$$

It was found that the initial rate of nitrosamide group accumulation is proportional to $[NO_2]^n$, where n ≈ 2.

As was shown by ESR, the reaction of NO, with N–H bonds also produces acylalkylnitroxyl macroradicals:

$$\text{~CONHCH}_2\text{~} + NO_2 \xrightarrow[HNO_2]{} \text{CON}^{\bullet}\text{CH}_2\text{~} \xrightarrow{NO_2} \text{~CON(ONO)CH}_2\text{~} \qquad (33)$$

$$\text{~CON(ONO)CH}_2\text{~} \rightleftharpoons \text{~CON(O}^{\bullet}\text{)CH}_2\text{~} + NO \qquad (34)$$

As well as in PCA, the interaction of NO_2 with PVP leads to UV bands characteristic of the nitrosamide group [20]. The formation of these groups in PVP is associated with splitting of the side-chain cyclic fragments from the main chain:

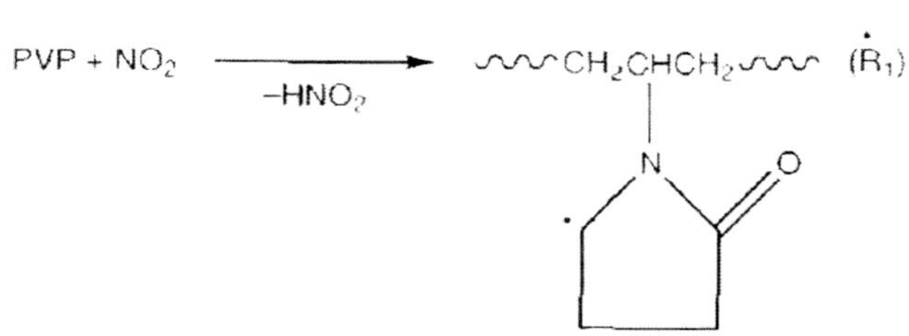

(35)

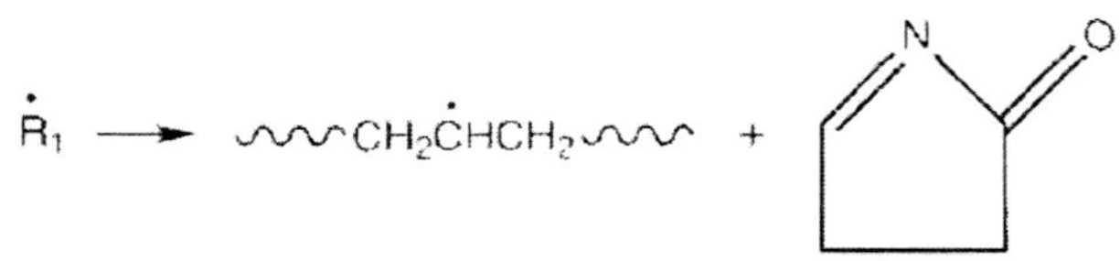

(36)

Thereafter, the reaction of NO_2 with the cyclic double bond gives rise to the nitrosamide product:

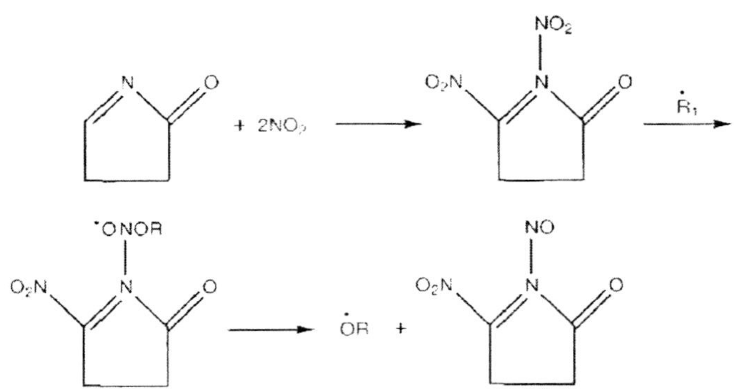

(37)

The ESR spectra observed when NO_2 (10^{-4} to 10^{-3} mol/l) reacts with PVP represent the superposition of the signals of acylalkylnitroxyl radicals ($A^N_{\parallel}$ = 1.94 mT, $g_{\parallel}$ = 2.003) and iminoxyl radicals ($A^N_{\perp}$ = 4.33 mT, $A^N_{\parallel}$ = 2.44 mT, $g_{\parallel}$ = 2.0029, $g_{\perp}$ = 2.0053). The formation of these iminoxyl radicals is initiated by the hydrogen atom abstraction reaction from C–H bonds that are in the a-position with respect to the amide group by reaction (35) and the following reaction:

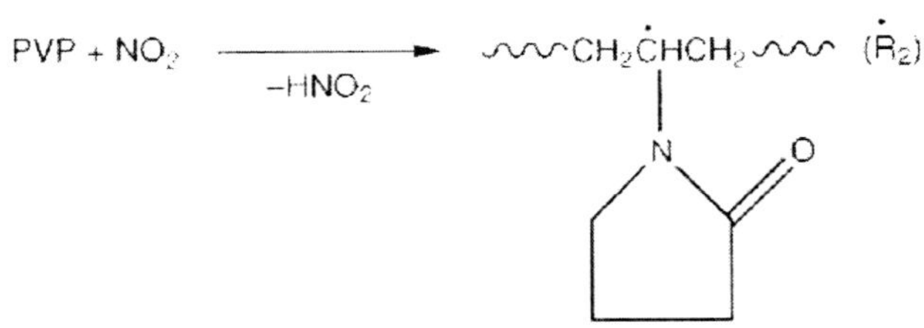

(38)

Nitric acid is thought to be the source of nitrogen oxide in the given system:

$$2HNO_2 \rightarrow H_2O + NO_2 + NO \quad (39)$$

The recombination of NO and $R^{\bullet}_1$ initiates the formation of iminoxyl radicals:

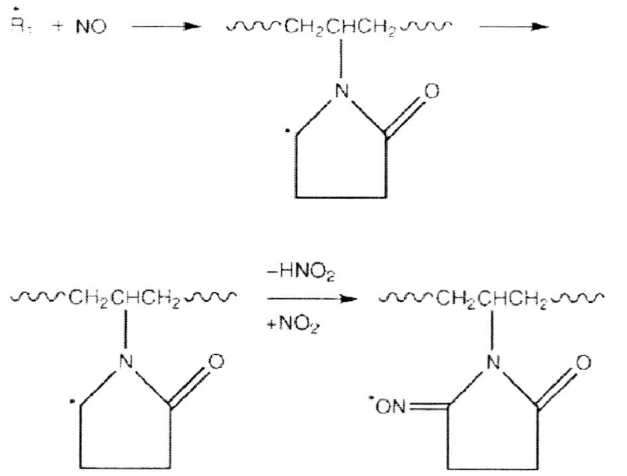

(40)

The formation of NO explains the production of acylalkylnitroxyl radicals as follows:

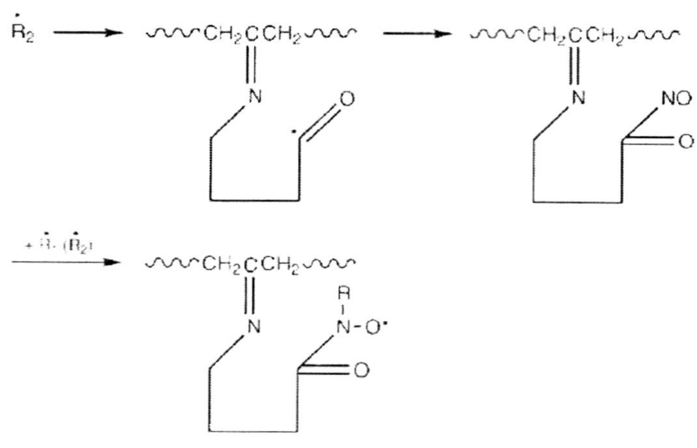

(41)

An approach based on the analysis of the composition of nitrogen-containing radicals in PVP depending on the content of filler aerosil has been put forward to elucidate the effect of polymer structural-physical organisation [21]. The influence of structural organisation may be manifest in the rates of iminoxyl and acylalkylnitroxyl radical formation. Filling gives the possibility of changing the physical structure of the polymer in interface layers. The decrease in the molecular packing density as a result of filling can accelerate the rate of reaction (41) involving breakage of the pyrrolidone cycle. The packing density decrease enhances the reaction rate through the promotion of mutual diffusion of $R^{\bullet}$ macroradicals and nitroso compounds. It is well established that the quantitative relation between iminoxyl and acylalkylnitroxyl radicals is changed with the degree of filling.

Formation of a gel fraction has been detected on exposure of polyurethane films to NO_2 [21]. Degradation of macromolecules simultaneously takes place in the sol fraction of the samples. The changes in the destruction degree and the gel-fraction yield with time are complex to analyse. The gel fraction at 333 K and $P(NO_2) = 20$ mm Hg initially increases up to 20% and thereafter reduces to nearly zero. The number of scissions in the sol fraction increases at the beginning, subsequently reduces, and then grows again. The exposure of films to NO_2 is accompanied by the release of CO_2 at all temperatures. The IR spectra in this case show N-H bond (3300 cm^{-1}) consumption. The proposed mechanism includes the reaction of NO_2 with the N-H groups of both the main chain and the side branches:

$$\text{\sim OCO-NH-CH}_2\text{\sim} + NO_2 \rightarrow \text{\sim OCO-N}^\bullet\text{-CH}_2\text{\sim} + HNO_2 \quad (42)$$

$$\text{\sim OCO-N(RH)-CH}_2\text{\sim} + NO_2 \rightarrow \text{\sim OCO-N(R}^\bullet\text{)-CH}_2\text{\sim} + HNO_2 \quad (43)$$

The recombination of ~OCO-$N^\bullet$-CH_2~ ($R^\bullet_1$) and ~OCO-N($R^\bullet$)-CH_2~ ($R^\bullet_2$) results in polymer crosslinking. The conversion of $R^\bullet_1$ causes macromolecule decomposition and CO_2 release. The exposure of polyurethane films to an NO_2 atmosphere or a mixture of NO_2, with air leads to the progressive reduction of the tensile strength limit [22].

The influence of NO_2 on the mechanical properties of polyamidoimide films has been considered at 323 K and $F(NO_2) = 13$ kPa [23]. The temperature dependences of the storage modulus E' and loss modulus E'' have been obtained for various times of NO_2 exposure. A nonmonotonic decrease of E' was observed at 473 K, but the maximum of the E' temperature dependence appears at approximately the same temperature. Samples exposed to NO_2 for eight days show an increase in E' at the glass transition temperature (563 K). The phenomenon is associated with chain breakage and the recombination of macroradicals giving rise to crosslinking. Chain breakage is supported by results obtained by the present authors. The ESR spectra of polyamidoimide exposed to an NO_2 atmosphere show the formation of iminoxyl radicals with spectral parameters that are close to those of PVP iminoxyl radicals. The possible mechanism of their formation includes the main-chain decomposition step as follows:

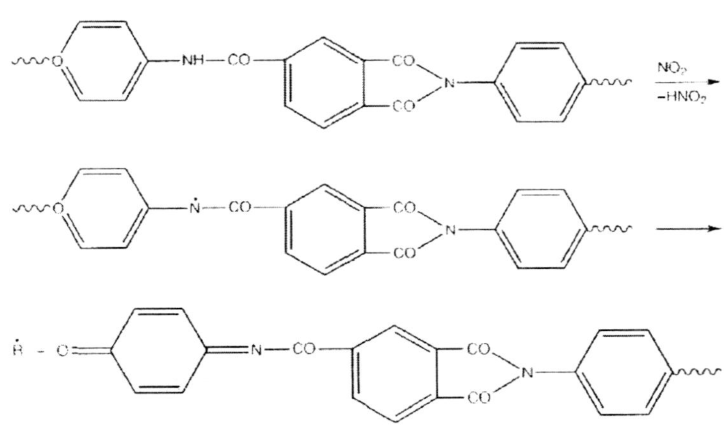

(44)

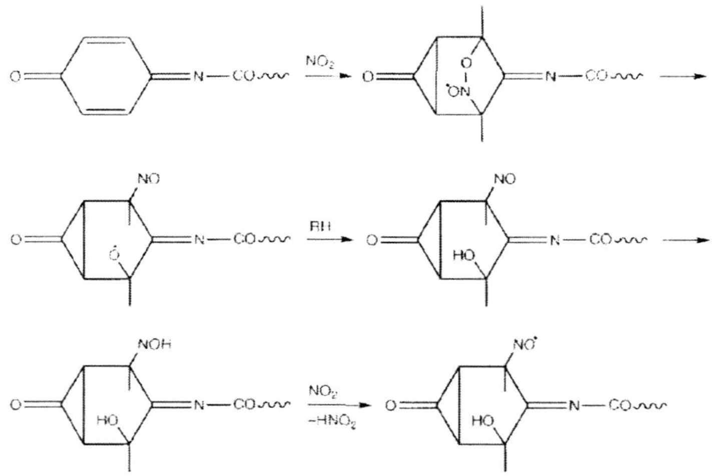

(45)

3. Reaction of Nitric Oxide with Polymers

Nitric oxide is a low-activity free radical and can be used as a 'counter' of radicals in gas and liquid phases. The reactions of alkyl radicals with NO lead to the formation of nitroso compounds, which are spin traps. Thus, the initiation of free-radical reactions in solid polymers in the presence of nitric oxide provides further information on their mechanism. It is well established that at room temperature NO is not able to remove allylic and tertiary hydrogen atoms and add to isolated double bonds [24-26]. There are discrepant opinions on the capability of NO to react with low molecular weight (low molar mass) dienes and polyenes. Some authors believe that NO is able to add to dienes and polyenes, for example, to substituted *o*-quinonedimethane, phorone and β-carotene, with the formation of free radicals [27-29]. Another way of looking at these reactions lies in the fact that they can be initiated by NO_2 impurities [25, 26].

This section of the review is concerned with radical reactions in polymers, induced by photo- and γ-irradiation, in the presence of nitric oxide. Irradiation of powdered PMMA in an NO atmosphere by the light of a mercury lamp results in the formation of three types-of macromolecuiar nitroxyl radicals [30]. The radical composition depends on temperature and the wavelength of the light. If the photolysis of PMMA is performed at room temperature using unfiltered light from a high-pressure mercury lamp, acylalkylnitroxyl radicals $R_1N(O^\bullet)C(=O)R_2$ are formed. The irradiation of samples at 383 K produces, in addition to acylalkylnitroxyl radicals, dialkylnitroxyl macroradicals $R_1N(O^\bullet)R_2$. Finally, if PMMA irradiation is carried out at room temperature using UV light with 260 nm < λ < 400 nm, the signal of iminoxyl radicals $R_1C(=NO^\bullet)R_2$ is also observed in the ESR spectrum.

Acetyl cellulose (AC) under action of light at room temperature gives rise to dialkyl- and acylalkylnitroxyl radicals [30]. The removal of NO from the samples leads to increasing of components of acylalkylnitroxyl radicals in the ESR spectrum. This phenomenon is probably connected with the formation of diamagnetic complexes of NO with acylalkylnitroxyl radicals. Dialkylnitroxyl radicals do not form complexes of this type at 298 K.

The γ-irradiation of PMMA at room temperature, as a photolysis, brings about the formation of acylalkylnitroxyl radicals [30]. Iminoxyl radicals also arise, but their quantity is essentially smaller than in AC under γ-irradiation.

The formation of nitroxyl radicals during photolysis as well as in the course of radiolysis of PMMA and AC in the presence of NO is explained by the following scheme:

$$\text{polymer} \xrightarrow{\gamma,\, h\nu} R_1^{\bullet}\ (R_2^{\bullet}) \qquad (46)$$

$$R_1^{\bullet} + NO \rightarrow R_1NO \qquad (47)$$

$$R_2^{\bullet} + R_1NO \rightarrow R_1N(^{\bullet}O)R_2 \qquad (48)$$

The structure of nitroxyl radicals is determined by the nature of the free radicals that are generated by γ- and photo-irradiation of PMMA and AC. Photo-irradiation of PMMA and AC leads to the formation of $^{\bullet}C(O)OCH_3$ radicals, which give in turn acylalkylnitroxyl radicals by reactions (46)-(48). Dialkylnitroxyl radicals arise when two macroradicals are involved in the reactions with NO.

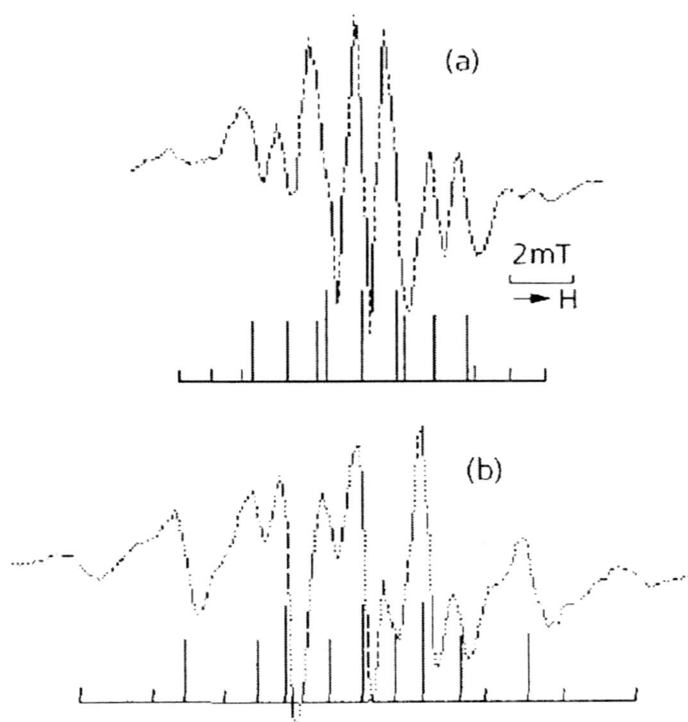

Figure 3. ESR spectra of perfluoronitroxyl radicals in PTFE films stretched to fourfold increase in its length at parallel (a) and perpendicular (b) orientation of magnetic field directions.

The free-radical reactions in solid polymers in the presence of NO are of particular significance for the preparation of spin-labelled polymers. This method has become particularly important for chemically inert, rigid and insoluble polymers, for instance, polytetrafluoroethylene (PTFE), because of the difficult problem of introducing spin labels by chemical reactions of nitroso compounds, nitrons or nitroxyl biradicals [31]. Oriented PTFE films γ-irradiated at room temperature in air after prolonged NO exposure contain nitroxyl radicals whose ESR spectra are displayed in Figure 3 [32].

The rotation of the samples leads to changes in angle a between the magnetic field and stretching directions. At 298 K and $\alpha = 0°$, the ESR spectrum is a triplet consisting of quintets with splitting of $A^N = 0.46$ mT and $A^F = 1.11$ mT, and $g_{||} = 2.0060$. At $\alpha = 90°$, the splittings increase to $A^N = 1.12$ mT and $A^F = 1.61$ mT, and $g_{\perp} = 2.0071$. The radicals observed are nitroxyl radicals with the following structure: $\sim CF_2\text{-}N(O^\bullet)\text{-}CF_2\sim$. A possible mechanism for nitroxyl macroradical synthesis has been suggested [32]. In an oxygen-containing atmosphere, some of the middle alkyl radicals formed in the course of "/-irradiation are capable of decomposing with rupture of the main chain as a result of the high energy transfer to these radicals:

$$\sim CF_2\text{–}CF_2\text{–}C^\bullet F\text{–}CF_2\sim \;\rightarrow\; \sim CF_2^\bullet + CF_2\text{=}CF\text{–}CF_2\sim \tag{49}$$

In the presence of oxygen, the terminal alkyl macroradicals can be oxidised to form terminal peroxy radicals:

$$\sim C^\bullet F_2 + O_2 \rightarrow \sim CF_2OO^\bullet \tag{50}$$

Under the action of NO on samples containing neighbouring terminal double bonds and peroxy radicals, the latter are converted into macromolecular nitrates and nitrites:

$$\sim CF_2\text{–}CF_2OO^\bullet + NO \longrightarrow [\sim CF_2OONO]\ldots \tag{51}$$

$$\longrightarrow [\sim CF_2O^\bullet + NO_2] \rightleftarrows \sim CF_2ONO_2; \quad [\sim CF_2O^\bullet + NO_2] \rightarrow \sim CF_2O^\bullet + NO_2 \tag{52}$$

$$\sim CF_2O^\bullet + NO \rightleftharpoons \sim CF_2ONO \tag{53}$$

Decomposition of alkoxy radicals in an NO atmosphere causes the synthesis of terminal nitroso compounds:

$$NO + {}^{\bullet}CF_2\text{–}CF_2\text{~} \rightarrow ON\text{–}CF_2\text{–}CF_2\text{~} \quad (57)$$

$$\text{~}CF_2\text{–}C(CF_3){=}CF_2 + ON\text{–}CF_2\text{–}CF_2\text{~} \rightarrow \text{~}CF_2\text{–}C^{\bullet}(CF_3)\text{–}CF_2\text{–}N(O^{\bullet})\text{–}CF_2\text{~} \rightarrow$$
$$\xrightarrow{+X} \text{~}CF_2\text{–}C(CF_3)(X)\text{–}CF_2\text{–}N(O^{\bullet})\text{–}CF_2\text{~} \quad (58)$$

where X is NO or NO_2. Nitrogen dioxide can be formed by the interaction of NO with $RO_2^{\bullet}$ in reaction (51).

One more type of nitroxyl macroradical is observed if a powdered TFE-HFP, γ-irradiated in air and exposed to NO with subsequent evacuation, is subjected to light irradiation at $\lambda >$ 260 nm at 298 K [34]. In this case, a new type of nitroxyl macroradical with the structure $\text{~}CF_2\text{-}N(O^{\bullet})\text{-}CF_3$ was registered. The following scheme provides an explanation for the radical formation in TFE-HFP under the action of light:

$$R_cO_2^{\bullet} + NO \rightarrow [R_cOONO] \rightarrow R_cONO_2 \quad (59)$$

$$R_eO_2^{\bullet} + NO \rightarrow [R_eOONO] \rightarrow R_eONO_2 \quad (60)$$

$$R_i^{\bullet} + NO \rightarrow R_iNO \quad (61)$$

$$R_cONO_2 + h\nu \rightarrow R_cO^{\bullet} + NO_2 \quad (62)$$

$$R_eONO_2 + h\nu \rightarrow R_eO^{\bullet} + NO_2 \quad (63)$$

$$R_iNO + h\nu \rightarrow R_i^{\bullet} + NO \quad (64)$$

$$R_cO^{\bullet} \rightarrow \text{~}CF_2\text{–}CFO + {}^{\bullet}CF_2\text{–}CF_2\text{~} \quad (65)$$

$$R_eO^{\bullet} \rightarrow \text{~}CF_2\text{–}CF_2^{\bullet} + CF_2O \quad (66)$$

$$\text{~}CF_2\text{–}CF_2^{\bullet} + NO \rightarrow \text{~}CF_2\text{–}CF_2\text{–}NO \quad (67)$$

$$\text{~}CF_2\text{–}C^{\bullet}(CF_3)\text{–}CF_2\text{~} + NO_2 \rightarrow \text{~}CF_2\text{–}C(CF_3)(ONO)\text{–}CF_2\text{~} \quad (68)$$

$$\sim CF_2\text{–}C(CF_3)(ONO)\text{–}CF_2\sim \; + \; h\nu \; \rightarrow \; \sim CF_2\text{–}C(CF_3)(O^{\bullet})\text{–}CF_2\sim \; + \; NO \qquad (69)$$

$$\sim CF_2\text{–}C(CF_3)(O^{\bullet})\text{–}CF_2\sim \; \rightarrow \; {}^{\bullet}CF_3 \; + \; \sim CF_2\text{–}C(=O)\text{–}CF_2\sim \qquad (70)$$

$${}^{\bullet}CF_3 \; + \; NO \; \rightarrow \; CF_3NO \qquad (71)$$

$${}^{\bullet}CF_3 \; + \; ON\text{–}CF_2\text{–}CF_2\sim \; \rightarrow \; CF_3\text{–}N(O^{\bullet})\text{–}CF_2\text{–}CF_2\sim \qquad (72)$$

$$CF_3NO \; + \; {}^{\bullet}CF_2\text{–}CF_2\sim \; \rightarrow \; CF_3\text{–}N(O^{\bullet})\text{–}CF_2\text{–}CF_2\sim \qquad (73)$$

It is obvious that the simultaneous action of light and NO on TFE-HFP results in macromolecular decomposition.

Polymer hydroperoxides are active participants in degradation processes. The reactions of nitrogen oxides with these particles are of interest to understand the mechanism of the influence of pollutants on polymer stability in the course of the oxidation process. The phenomenon of hydroperoxide decomposition under the action of NO was discussed long ago using both macromolecular peroxides and their low molecular weight analogues [34]. Some authors assumed that the primary stage of peroxide decomposition can be represented by the reaction [35]:

$$ROOH \; + \; NO \; \rightarrow \; RO^{\bullet} \; + \; HONO \qquad (74)$$

Another mechanism [36] suggests that the reaction proceeds with the formation of peroxide radicals:

$$ROOH \; + \; NO \; \rightarrow \; ROO^{\bullet} \; + \; HNO \qquad (75)$$

The kinetics of hydroperoxide decomposition in PP at 298 K and various partial pressures of NO has been studied in detail 134]. The decomposition kinetics are shown in Figure 4.

As can be seen, the hydroperoxide consumption rate is initially low and then sharply increases. The observed character of the kinetic curves cannot be explained by reactions (74) or (75). According to the ESR data, the decomposition of PP hydroperoxide in an NO atmosphere gives dialkylnitroxyl radicals. It was shown that the induction periods for the hydroperoxide decomposition and nitroxyl radical accumulation are very sensitive to the presence of trace amounts of higher nitrogen oxides. This leads to the conclusion that the interaction of hydroperoxide with NO is more likely to proceed according to the scheme:

$$ROOH + N_2O_3 \rightarrow [ROONO] + HNO_2 \qquad (76)$$

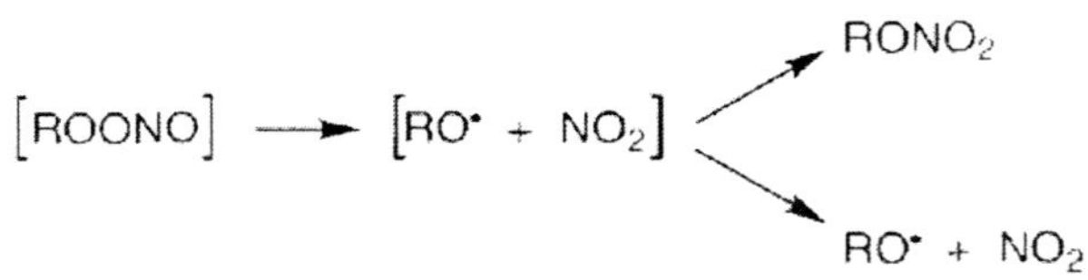

(77, 78)

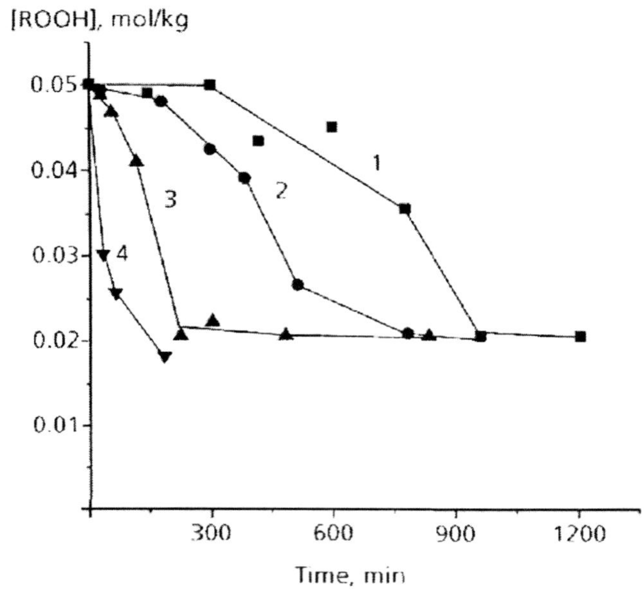

Figure 4. Kinetics of PP hydroperoxide decomposition in NO at various concentrations (1-3) and NO + NO_2 mixture (4); (1) 1.61×10^{-3}, (2) 3.22×10^{-3}, (3) 4.13×10^{-3}, (4) 3.1×10^{-3} NO and 3.0×10^{-6} NO_2 mol/l.

Alkoxy radicals may decompose or enter into substitution reactions with macromolecules to form chain $R_c^\bullet$ and end $R_e^\bullet$ alkyl macroradicals, and low molecular weight alkyl radicals $r^\bullet$, which with NO give nitroso compounds:

$$RO^\bullet \rightarrow R_c^\bullet\ (R_e^\bullet,\ r^\bullet) \qquad (79)$$

$$R_c^\bullet\ (R_e^\bullet,\ r^\bullet) + NO \rightarrow R_cNO\ (R_eNO,\ rNO) \qquad (80)$$

The increase in the rate of hydroperoxide decomposition with time can be related to reactions proceeding with participation of such nitroso compounds:

$$r'OOH + r''NO \rightarrow r'O^\bullet + r'\text{–}N(O^\bullet)\text{–}OH \qquad (81)$$

$$r''\text{-}N(O^{\bullet})\text{-}OH \rightarrow r''^{\bullet} + HNO_2 \tag{82}$$

The alkyl radicals formed in the system may stimulate hydroperoxide decomposition [37]:

$$r^{\bullet}\ (R_c^{\bullet},\ R_e^{\bullet}) + ROOH \rightarrow rH\ (R_cH,\ R_eH) + RO_2^{\bullet} \tag{83}$$

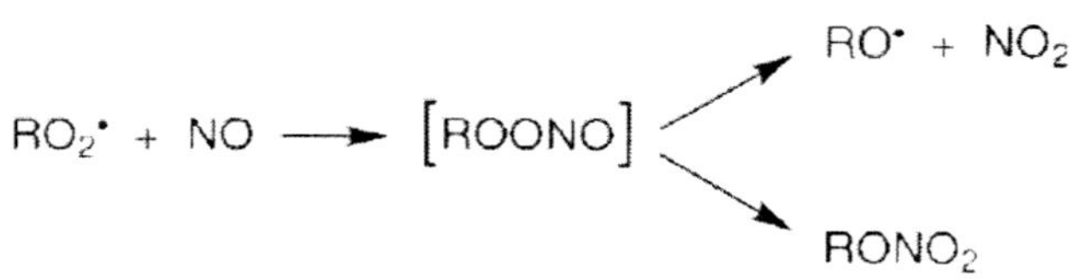

(84, 85)

Another process that can increase the hydroperoxide decomposition rate is the disproportionation of NO to N_2 and $NO_3^{\bullet}$ with the participation of nitroso compounds [24]:

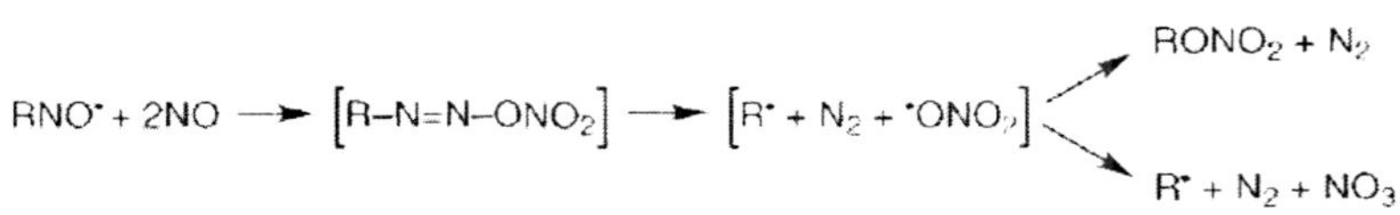

$$NO + NO_3^{\bullet} \rightarrow 2NO_2$$

(86, 87)

Reactions (83)-(88) may lead to an increasing NO_2 concentration in the system and, consequently, result in the acceleration of reaction (76).

4. Conclusion

Nitrogen oxides are capable of influencing the free-radical stages of polymeric material aging in polluted atmospheres. Nitric oxide is a comparatively low-activity free radical, and it cannot abstract even labile hydrogen atoms at ordinary temperatures to initiate the radical degradation process. On the other hand, NO effectively recombines with free radicals. This reaction is apparently controlled in solid polymers by the gas diffusion rate, and NO is capable of terminating the oxidation chain by reaction with peroxy and alkyl macroradicals. The reaction of NO with alkyl radicals gives nitroso compounds, which are spin traps accepting free radicals. This process can slow down polymer degradation in the presence of nitrogen oxides in subsequent conversions, which can break down into alkoxy radicals, effecting the degradation of macromolecules. In addition, nitric oxide initiates the decomposition of bydroperoxides resulting from oxidation of polymers.

Nitrogen dioxide is a more active free radical as compared with NO, and is able to break off the labile hydrogen atoms at room temperature as well as to add to the C=C bonds of macromolecules, inducing free-radical degradation of polymers. At the same time, the NO_2 radical can inhibit the free-radical reactions giving nitrogen-containing molecules by the reactions with alkyl, alkoxy and peroxy radicals. The thermal and photochemical conversions of these products also affect the aging process of polymeric materials. Nitrogen dioxide is an initiator of the free-radical degradation of polyolefins at elevated temperatures.

The low stability of polyamides to the action of NO_2 is quite surprising, because the N-H bond of the amide group is rather strong. Therefore, the mechanism of polyamide degradation connected with hydrogen atom abstraction by NO_2 from N-H bonds is not fully elucidated.

REFERENCES

[1] H.H.G. Jellinek in *Aspects of Degradation and Stabilisation of Polymers*, Ed., H.H.G. Jellinek, Elsevier, Amsterdam, The Netherlands, 19`8, Chapter 9.

[2] N.M. Emanuel and A.L. Buchachenko, *Chemical Physics of Polymer Degradation and Stabilisation,* VNU Science Press, Utrecht, The Netherlands, 1987.

[3] Karpukhin, *Usppekhi Khimii,* 1978, 47, 6, 1119.

[4] T.C. Hall and F.E. Blacet, *Journal of Chemical Physics,* 1952, 20, 11, 1745.

[5] J.G. Calvert and J.N. Pitts Jr., *Photochemistry,* John Wiley, New York, NY, USA, 1966.

[6] H.H.G. Jellinek, F. Flajsman and F.J. Kryman, *Journal of Applied Polymer Science,* 1969,13, 1, 107.

[7] T. Ogihara, *Bulletin of the Chemical Society of Japan,* 1963, 36, 1, 58.

[8] T. Ogihara, S. Tsuchiya and K. Kuratani, *Bulletin of the Chemical Society of Japan,* 1965,38,6,978.

[9] H.H.G. Jellinek and Y Toyoshima, *Journal of Polymer Science, Part A-l: Polymer Chemistry,* 1967, 5, 12, 3214.

[10] H.H.G. Jellinek and F. Flajsman, *Journal of Polymer Science, Part A~l: Polymer Chemistry,* 1969,7,4, 1153.

[11] A. Huber, *Einflufi von Schwefeldioxid und Stickstoffdioxid auf Polymere in Luft unter Belichtung,* University of Stuttgart, Germany, 1988, 187. [Ph.D Thesis].

[12] H.H.G. Jellinek and F. *Flajsman, Journal of Polymer Science, Part A-l: Polymer Chemistry,* 1970,8,3,711.

[13] H.H.G. Jellinek and P. Hrdlovic, *Journal of Polymer Science, Part A-l: Polymer Chemistry,* 1971,9,5, 1219.

[14] TV. Pokholok and G.B. Pariiskii, *Polymer Science, Series A,* 1997, 39, 7, 765.

[15] E.N. Degtyarev, TV. Pokholok, G.B. Pariiskii and O.E. Yakimchenko, *Zhurnal Fizicheskoi Khimii,* 1994, 68, 3, 461.

[16] T.V. Pokholok, G.B. Pariiskii and G.O. Bragina, *Vysokomolekulyarnye Soedineniya, Seriya A,* 1989, 31, 10, 2049.

[17] H.H.G. Jellinek and A. Chaudhuri, *Journal of Polymer Science, Part A-l: Polymer Chemistry,* 1972, 10, 6, 1773.

[18] H.H.G. Jellinek, R. Yokota and Y. Itoh, *Polymer Journal,* 1973, 4, 6, 601.

[19] H. Herzlinger, B. Kuster and H. Essig, *Textile Praxis International,* 1989, 44, 6, 574,655,661.

[20] I.S. Gaponova, E.Y. Davydov, G.G. Makarov, G.B. Pariiskii and V.P. Pustoshnyi, *Polymer Science, Series A,* 1998, 40, 4, 309.
[21] H.H.G. jellinek and T.J.Y. Wang, *Journal of Polymer Science, Polymer Chemistry Edition,* 1973,11, 12,3227.
[22] H.H.G. Jellinek, F. Martin and J. Wegener, *Journal of Applied Polymer Science,* 1974,18,6, 1773.
[23] H. Kambe and R. Yokota, *Proceedings of the 2nd International Symposium on Degradation and Stabilisation of Polymers,* Dubrovnik, Yugoslavia, 1978, Paper No.39.
[24] J.F. Brown, Jr., *Journal of the American Chemical Society,* 1957, 79, 10, 2480.
[25] A. Rockenbauer and L. Korecz, *Chemical Communications,* 1994, 145.
[26] J.S.B. Park and J.C. Walton, *Perkin Transactions 2,* 1997, 12, 2579.
[27] H.-G. Korth, R. Sustmann, P. Lommes, T. Paul, A. Ernst, H. de Groot, L. Hughes and K.U. Ingold, *Journal of the American Chemical Society,* 1994, 116, 7, 2767.
[28] Gabr and M.C.R. Symons, *Faraday Transactions,* 1996, 92, 10, 1767.
[29] Gabr, R.P. Patel, M.C.R. Symons and M.T. Wilson, *Chemical Communications,* 1995,9,915.
[30] I.S. Gaponova, G.B. Pariiskii and D.Ya. Toptygin, *Vysokomolekulyarnye Soedineniya, Seriya* A, 1988, 30, 2, 262.
[31] A.M. Wasserman and A.L. Kovarskii, *Spinovye Metki i Zondy v Fizikokhimii Polimerov (Spin Labels and Probes in Physical Chemistry of Polymers),* Nauka, Moscow, Russia, 1986.
[32] LS. Gaponova and G.B. Pariiskii, *Chemical Physics Reports,* 1997, 16, 10, 1795
[33] LS. Gaponova and G.B. Pariiskii, *Polymer Science, Series B,* 1998, 40, 11-12, 394.
[34] LS. Gaponova and G.B. Pariiskii, *Polymer Science, Series* A, *1995,* 37, 11, 1133.
[35] J.R. Shelton and R.F. Kopczewski, *Journal of Organic Chemistry,* 1967, 32, 9, 2908.
[36] DJ. Carlsson, R. Brousseau, C. Zhang and D.M. Wiles, *Polymer Degradation and Stability,* 1987, 17, 4, 303.
[37] K. Ingold and B. Roberts, *Free-Radical Substitution Reactions,* John Wiley, New York, NY, USA, 1972.

In: Order and Disorder in Polymer Reactivity
Editors: G. E. Zaikov and B. A. Howell, pp. 77-113
ISBN 1-60021-145-3

Chapter 7

Polymer Materials with The Structural Inhomogeneities for Modern Optical Devices

***N. Lekishvili*[1], *L. Nadareishvili*[1] *and G. Zaikov*[2]**

[1] I. Javakhishvili Tbilisi State University, Faculty of Chemistry, the Scientific Center of Nontraditional Materials; 1, Ilia Chavchavadze ave., 0128 Tbilisi, Georgia

[2] N.E. Emanuel Institute of Biochemical Physics of the Academy of Sciences of Russia 119991 Moscow, 5, N.N. Kosigin street, Russian Federation

Abstract

Polymer media with the structural in-homogeneity, determining of the gradient of their optical properties are described. The general tendencies of development of gradient optics are discussed. The results of novel scientific researches on polymer gradient light-focusing areas (elements) are generalized and analyzed. The dependences of the optical and other properties of gradient elements on various factors (monomers nature, the degree of their conversion, temperature, durability of the gradient layer, formation, etc.) on the control and form of the gradient element production technology are discussed. The new methods of the producing of gradient areas from solid substances are described.

Introduction

The future of microelectronics, integral and fiber optics is closely connected with creation of optically transparent polymer materials with a gradient of optical and other properties and how they behave over long periods of time. These new materials optics – gradient or GRIN-optics (GRIN-gradient of refractive index) arises from the possession by these materials of a gradient of refractive index within the material. The creation of these materials has allowed radical improvements to be made of existing optical devises. The manufacture of these devises is impossible on the bases of traditional homogeneous optical materials. Unlike of traditional optics, the trajectory of the light ray in the gradient medium is continuously nonlinear, which leads to deviation of the light ray, and in particular, to radial-symmetric profile of refractive index distribution and ray focusing [1, 2]. GRIN-elements are capable independently, without additional means, to form and broadcast the image. Among these elements the ones with desirable distribution of refractive index from periphery to center - selfocs have a special place. They are used in the communication optical systems, copy-many folding techniques, optical devises, etc. [3-17].

The development of gradient optics creates the possibility of polymer materials successfully competing with traditional optical materials - inorganic glasses, and in some cases to surpass them. The methods described enable the use of a relatively wide range of polymer types, for example, polycondensational ones, also, nontraditional (for gradient optics) physical factors (magnetic and inhomogeneous electric fields, Laser irradiation, etc.) for the creation of optical gradient properties. Besides it also necessary to increase of the value of refractive index and create polymer gradient elements with high environmental stability.

The other new and practically more interesting direction of gradient optics is the creation of optical materials with the gradient of birefringence or GB-optics [17]. Early, it was not possible to manufacture polymer GB-elements (the elements with the gradient of birefringence). Polymer elements with the gradient of birefringence have great potential practical application in the optical device manufacture [18]. The production of polymer materials with a given axial (may be radial) gradient of birefringence were achieved using a mechanical stress with a definite inhomogeneity, superposed on the optical transparent polymer film samples with the desirable axial (may be radial) gradient and degree of elongation [17]. The present review will be focused on the following basic topics:

- Improvement of optical characteristics polymeric GRIN materials due to upgrade of the manufacturing methods used for their production;
- Expansion of the classes of used polymers and also modified/synthesized new polymers;
- Development of essentially new methods of GB-elements manufactures which upgrades the existing methods of GRIN-elements for use in special optical devices;
- Creation of inhomogeneous gradient layers on surfaces of spherical lenses with the aim of geometrical aberration corrections;
- Research on the possible creation of polymeric media with gradient of optical properties (gradient of refraction index, gradient of birefringence) by use of some, non-conventional in gradient optics in-homogeneity physical factors;
- Investigation of gradient deformation, as new kind of polymeric bodies deformation (research of feature gradient deformation influence on orientation properties of polymers; investigation of topography of deformation distribution in polymeric samples in conditions of gradient loading);
- Search for new applications for the use of GRIN-elements and GB-elements, including applied and polarization optics, optical instrument making industry, etc;

1. THE GENERAL TENDENCIES OF DEVELOPMENT OF POLYMER GRADIENT OPTICS

The intensive investigations on gradient optics started about 30-40 years ago. The research started with the investigation of the behavior of electromagnetic waves in inhomogeneity medium, particularly, in the medium with a refractive index gradient. In the 1960-70, cylindrical elements with a parabolic distribution of refractive index capable of focusing light rays and transfer of pictures were created [12]. The first results on polymer selfocs were published at beginning of the 1970's. These GRIN-elements were based on the monomers – diallylisophtalate (DAIP) and methylmethacrylate (MMA). Earlier researches [3, 5, 6, 8-12] used the method of exchange molecular diffusion and photopolymerization of two or three functional allyl and vinyl monomers with various reaction capabilities and different value of refractive index [20-23]. On photopolymerization, the gradient formation factors present different relative activity ($r_{i,i}$) and various values of refractive indexes ($n_{1,2}$) of initial monomers (for example, N-vinylcarbazole (N-VK) - MMA, N-vinylphtalimid - MMA, N-VK - MMA-vinylacetate, vinyl-benzoate-acrilonitryl-MMA, etc.). For the creation of co-polymer gradient materials, it is necessary to place some conditions on the properties of the two monomers; M_1 and M_2, $r_1 < r_2$ and $n_1 > n_2$ (where $\boldsymbol{r_1}$ and $\boldsymbol{r_2}$ are constants of copolymerization and $\boldsymbol{n_1}$ and $\boldsymbol{n_2}$ are refractive indexes of M_1 and M_2. The use of three monomers places more rigid constraints on the selection [17]. Besides of some positive factors [one-stage process, experimental simplicity-copolymerization is carried out in hermetic closed quartz vessels which are rotated under influence of UV-irradiation; and the possibility of formation of polymer optical fibers (POF) with the gradient of refractive index] this method has such essential deficiencies, for example, the restricted number of corresponding monomers; the complexity of the control factors influencing the formation of a given distribution of refractive index. A distribution of refractive index (RID) in the corresponding optical elements leads to properties which are far from ideal; optical characteristics of selfocs, obtained by this method do not satisfied to presented technical requirements and it is impossible to reach to stable optical characteristics of prepared elements because of the high level (great amount) of residual (non-conversed) monomers. For these reasons the application of photo-polymerization method for the production of GRIN-elements is restricted, as compared with the diffusion method.

The selfocs on the basis of matrix (gel-polymers from monomers with high value of refractive index) from Diallylphtalate (DAIP) [25-27], diethylenglycolbisallylcarbonate (DEGBAC) [27-29], triallylcyanurate (TACU) [30], dimethacrilatediethylenglycol (DMEG) [31], diallylester of metacarborandicarbonacid (DAEMCA) [32], etc have been reported. The organic and element-oganic methacrylates [MMA, butylmethacrylate (BMA), silicon and fluorine-organic methacrylates], vinyl chloride, etc., were used as monomers with low value of refractive indexes (penetrants-M_D) [3, 27-34].

The selfocs based on the majority of above monomers, besides having some valuable properties (for example large aperture angle, relatively high thermal stability, etc.) have restricted length ($L \geq 150$ mm) and diameter ($d \geq 3$ mm), which restricted the sphere of their application [17].

The logical way of increasing of the gradient of refractive index and improvement of optical-mechanical characteristics of selfocs was to search for new monomer-diffusers ($\mathbf{M_D}$)

with lower values of refractive index and compatible with the well tested gradient optics monomers for the matrix (DAIP, DEGBAC, etc.) and alternatively the synthesis of new copolymer matrix from two or more monomers, capable to controlling the optical and physical-mechanical properties of GRIN-elements [17]. Increasing the gradient of refractive index (Δn) with a simultaneous decrease of the chromatic aberration and improvement of some physical and mechanical properties of selfocs was achieved using of perfluorinated methacrylates as monomers with low refractive indexes [35-37]. It was shown [35] that it is possible to obtain selfocs with small chromatic aberration (about 0.008) by selection of DEGBAC-1,1,3-trihydroperfluorinepropyl methacrylate and DEGBAC-1,1,5-trihydroperfluorinepropyl methacrylate as monomer pairs. By using of the same fluorine containing monomer-diffusers and various monomers for matrix there were obtained selfocs with $\Delta n \geq$ (0.07-0.08) and high thermal and moisture resistant properties and with good optical-mechanical characteristics keeping these properties for a long time [36,37]. The selfocs with relatively big diameters for that time ($d \approx 6$-8mm) were obtained by chemical modification of matrix from DAIP by it copolymerization with malein anhydride-MA (the plasticized effect of MA). [38].

In [31, 39, 40] was shown, that besides of the relatively expensive matrix monomers (DAIP, TACU, etc.), the more chief industrial monomers (e.g. styrene), which give the copolymers with high refractive index and elastic properties. At present of small amounts of scission agents of the type of DMEG [31], silicon-organic dimethacrylates [39], olygocarbonatedimethacrylate (OCM-2) (Table 1) [40], etc. can also be used. However this method did not lead to wide application in gradient optics because of the low gradient of refractive index and complexity of the production process of matrix component with reproductively optical and other characteristics. Flexible bar-like selfocs are described in [3, 31], where the improvement in flexibility was obtained by selection of corresponding components:

a. For the matrix - the monomers type of oligoethylenglycoldimethacrylates (with the given length of chain of the ethylenglycol fragment) or their copolymers with triallylcyanurate (TACU) [31];
b. Penetrant with mobile molecular chain [35], and by the creation of gel-polymer matrix with low degree of cross linking [3].

The manufacture of selfocs with variable focus distance is given in [41].

After this, other methods for formation of the gradient of distribution of refractive index were published. Particularly, division in the gravitation field [11], polymer-analogical conversion of polymer films [42], and dipole forced - diffusion of monomers with definite dipole moment into the polymer matrix under influence of inhomogeneous electric field with the given inhomogeneity [43, 44], etc. For some of these methods is not necessary to provide additional manipulation to fix the distribution of refractive index gradient obtained [1].

In the design of GRIN-elements to obtain a desirable flexible index is the most desirable property and this involves successful selection of initial monomers or polymers; their transformation to a material with the desirable distribution of the refractive gradient and achieving this in real optical elements [12, 17].

The following factors: the temperature, composition of initial monomer mixture, gradient of field of initiation, etc., are all important in the creation of the distribution of refractive index gradient in copolymer composites [1, 17]. Among these parameters is the selection of practical conditions used in the fabrication of the refractive index gradient profile (**RIGP**) [27, 45].

Table 1. The composition of OCM-2 – styrene mixture and their copolymerization conditions [40]

№	Monomers		n_D^{20} of monomers mixture	T,K	Time, min.	Content of gel-fraction in copolymer, %
	Styrene, мol. %	OCM -2, мol. %				
1	98,7	1,3	1,5403	353	50	11,0
2	95,9	4,1	1,5336	353	35	16,8
3	89,0	11,0	1,5260	353	20	18,0
4	80,0	20,0	1,5120	353	15	23,9
5	80,0	20,0	1,5120	333	40	22,1
6	63,8	36,2	1,4960	343	26	24,0

The classic method of exchange molecular diffusion gives rather good results for the production of GRIN-elements. The creation of selfocs by this method, as a rule, demands the realization of thee consequent process (stages) [12, 17]:

1. The production of gel-polymer matrix on the bases of monomer, consisting of usually two or three unsaturated radicals (allyl or vinyl):
2. The exchange molecular diffusion between residual monomer of gel-polymer matrix (monomer with high value of refractive index) and monomer-diffuser (monomer with more low value of refractive index relatively to matrix monomer);
3. The fixation process (thermo, photo, γ-radiation, etc.) of the gradient of concentration distribution of monomer-diffuser, and, respectively of profile of refractive index distribution.

The first stage which involves the creation of the gel-polymer matrix, and copolymerization of monomer-diffuser (penetrant) with matrix are typical complex hetero-phase processes, which are difficult to characterize and require a full arsenal of physical and chemical research methods to study. The factors controlling the formation of the gel-polymerization matrix with necessary content of gel-fraction, for creation of simple with corresponding geometry, have been established experimentally [45, 49]. By using of refractometric, spectroscopic (NMR and IR), electron paramagnetic resonance (EPR), differential-scanned calorimetric (DSC), thermo-mechanical and kinetic methods, the mechanism and structural peculiarity of formation of gel-polymer matrix have been studied. The saturated amounts of minimal conversion for formation of substance of space-bounded structure with necessary consistence and capable to keep of the given geometry of corresponding monomers (for DAIP-13,5 %, for DEGBAC-9,5 % and for DAEMCA-15-16 %) were established [39, 45]. It was shown, that application of above mentioned methods can define both the monomer content and gradual conversion of double bonds during full process of copolymerization and estimate the structural-morphological characteristics of forming polymer reflected on the properties of optical materials [12]. It was also shown, that effects exist which control the generation of the space cross-linked structure with definite physical and mechanical properties as a consequence of morphological defects [49]. The formation of these defects is determined by natural and technogenical factors and **is** often hard to control. The ways to control the gel-formation process and degree of cross linking of the gel-polymer matrix have been studied [27, 50]. Based on investigations using of selfoc model systems, the effect of different factors on the formation of gradient profile of refractive index at the stage both diffusion and fixation of **RIGP** have been studied [3, 6, 8, 13, 14, 51-60]. On the basis of analysis of the curves n=f(r) and n=f(r^2) (Fig.1), obtained for various cases, it was established the way in which various factors (temperature, initiator concentration, diffusion durability, etc.) influence the properties of the material, the general character of the **GPRI** is practically identical for one-type of monomer.

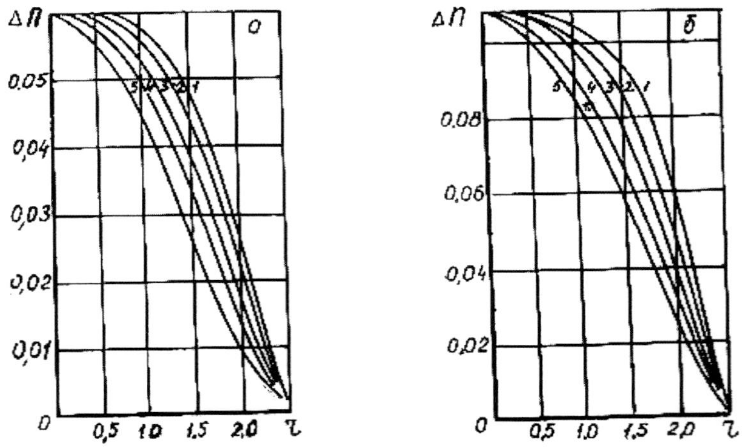

Figure 1. The refractive index distribution corves obtained in the models of the selfocs on the basis of monomer pairs: a) DEGBAC-4FMA, τ_D : 1-5 min., 2-8 min., 3-18 min., 4-24 min.,5-34 min.; b) DAIP-4FMA, τ_D : 1-5 min., 2-10 min., 3-15 min., 4-25 min.,5-35 min.

The quantitative characteristics of the process are different (Table 2) [39] and depend on the structure and molar volume of penetrants (monomer-diffusers) and on their ability to react, molecular mobility, extent of the degree of cross-linking of the matrix, diffusion process temperature and durability, initiator concentration and regime of fixation of RIGP [12].

It was established, that, in most cases, the process of exchange diffusion follows Fick law [60], however, for some monomer systems (e.g. for DAIP-MA-MMA) the diffusion is definitely anomalous and connected with an acceleration of the diffusion, as compared with the predictions of Fickian behavior and is influenced by increasing the diameter of the cylinder. [60].

To explain this phenomenon the ideas of a jump in the diffusion coefficient as a result of plastification of the copolymer based on DAIP-MA by MMA (monomer-diffuser) [60] has been proposed. It was observed [3, 17, 27, 39, 51, 52, 58], that in the resulting optical characteristics of selfocs that the composition of the reaction mixture at the gel point, the ratio of penetrant diffusion rate on the rate of it copolymerization with gel-polymer matrix are both important.

Table 2. The conditions of formation and main parameters of RIGP manufactured in the selfoc models on the bases of DAIP and DEGBAC at diffusion ($T_{dif.}$=323K) perflouorine-alkylmethacrylates [nF(M)A] [58]

Monomer of matrix*	Monomer-diffuser**	Durability of diffusion, *τ*, мин.	Gradient of refractive index, Δn	σ Δn , %	$\Delta n_{exp.}/\Delta n_{teor.}$, %
DAIP	4 FMA	35	0,1010	3,09	97
DAIP	8 FMA	54	0,1010	3,45	87
DAIP	12 FMA	89	---	---	20
DAIP	4 FMA	19	0,1120	3,03	97
DEGBAC	4 FMA	34	0,0530	3,77	98
DEGBAC	8 FMA	53	0,0576	6,10	87
DEGBAC	12 FMA	87	0,0180	---	25
DEGBAC	4 FA	46	0,0660	4,54	92

* DAIP- Diallylisophtalate; DEGBAC – diethyleneglycol-bis(allylcarbonate)

** n F(M)A: CH_2=C(CH_3)C(O)O CH_2 (CF_2 $)_n$ H, where n=2, 4, 6; 4FA: CH_2=CHC(O)O CH_2 (CF_2 $)_2$H;

Each of these processes for the production of the GRIN-elements is characterized by many parameters. They are connected to each to other and, optimization of the gradient elements is a very difficult problem. For a long time, the production of GRIN-elements by diffusion exchange method was founded on empirical recommendations. It is clear that, without the creation of a systematic approach, the essential problem remains, the reproduction of results even for one and the same monomer pair of monomers, the increasing discrimination ability of the polymer selfocs and obtaining gradient elements with unrestricted length and large diameter too [17].

The further development of diffusion technology of GRIN-elements required a more systematic approach to be adopted, both with regards the processes of generation of the matrix, formation of the given gradient of the refractive index and establishment of the efficient correlation between the created **RIGP** and content of the initial gradient composite, technological regime of production, physical-chemical and optical-mechanical characteristics of corresponding GRIN-elements.

In 1990's, physical and mathematical models describing of the characteristic of different refractive index gradient materials, in combination with investigation of optical characteristics of real systems, were published 57-60]. The possibility of creation of morphology and stratification of isorefraction in the GRIN-elements through diffusion process in the cylindrical segments were considered [61]. In the structural-anisotropic solids (e.g. gradient materials) there exists isostructures as layers of material with equal structure, at all "points" and within the limits of given layer the physical-chemical properties including the optical ones are the same. Such materials relate to ability to produce refraction of a light beam can be called isorefractive materials. So, that the formation of a refractive index gradient consists in the design of isorefraction layers with a definite morphology as stratified layers. In the case of the exchange diffusion method, the gel-polymer matrix may be consider as equilibrium system, consisting of stratified isoconcentration layers with a definite morphology, where in the each "point" of each separate layer the concentration of composites is one and same, but differ from other layers by it's concentration's. After cessation of the exchange diffusion process and establishment of difference in concentration, the stratified isoconcentration layers transform to stable isostructure medium-isorefractive layers. The morphology of isoconcentration and peculiarity of their stratification depend on such factors, as the relief of matrix surface, view of section, the regime of gradient formation process (isocronic / anisocronic), etc. [61].

There are three main types of gradient elements [18], which may be characterized in the terms of isorefractive layers. In the axial gradient elements the mutual parallel isorefractive layers are placed normally to optical axis-direction of gradient. At present of the spherical gradient the isoconcentrations have been viewed as concentric spheres, but in radial gradient elements are coaxial tubes. The penetration of diffuser-gradient former to the matrix takes place in the perpendicular direction relatively to matrix surface.

In the case of the spherical or cylindrical surface they present the radiuses, which crosses at one and same point. Inhomogenous gradient layer is formatted when at the process of creation of isotropic gradient (exchange diffusion) the directions of initial penetration of diffuser – gradient former in the matrix with plate (or spherical) surface and it further distribution in the volume of matrix do not coincidence. For example, if the flat matrix is covered by non-penetrating mask in which there is a hole (from which the monomer-diffuser penetrate into matrix), inhomogenous gradient layer is formatted simultaneously with axial and radial gradients [61].

At penetration of monomer-diffuser to the gel-polymer matrix along the radiuses of spherical (or cylindrical) surface the front of spreading of gradient-former is compressed, homogenous gradient layer is formed with parallel stratified isorefraction (in the view of concentrated sphere or coaxial tubes) (Figure 2).

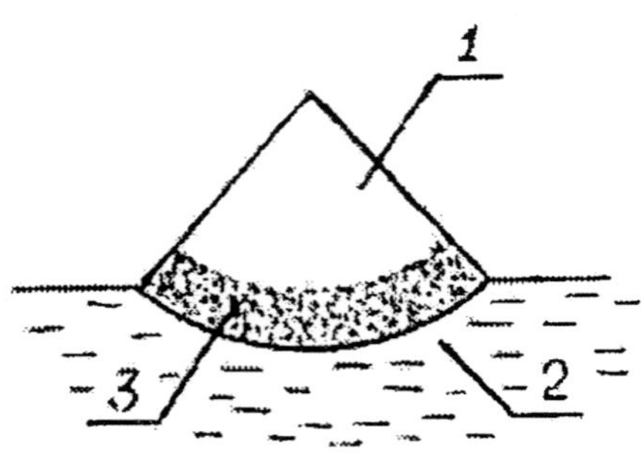

Figure 2. Scheme of generation of homogenous gradient layer in the spherical sector (radial section): 1.- spherical/cylindrical sector; 2.-diffusant; 3.- homogenous gradient layer.

However, if in the gradient-former introduce the foundation of the cylinder with spherical convexity (Figure 3, a) than may be formatted inhomogenous gradient layer; in the center of the foundation the contraction of the front of translation of the penetrant will be placed, but at the periphery, on the contrary-the widening; the penetrant will occupy additionally the space between imagine sphere sector 3 and the side surface of the cylinder 1 (Figure 3, position 2). Therefore, in this region and in the adjacent space will be damaged the isorefraction. It will be increased the radius of their curvature; the thickness of the gradient layer will be decreased – appears the inhomogenous gradient layer [61].

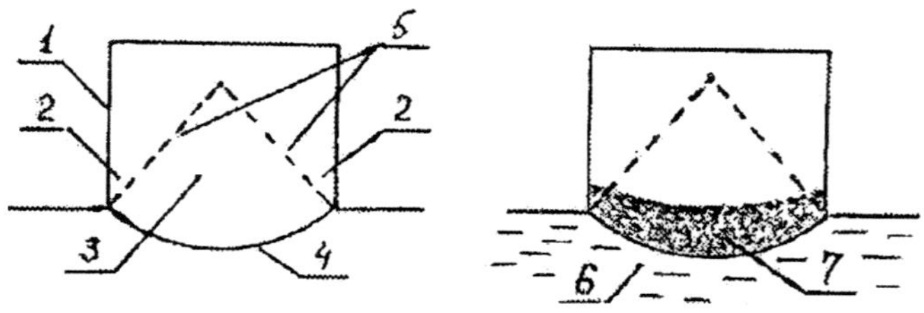

Figure 3. Diametric section of the gel-polymer cylindrical lens: a) the wall of cylinder (1); the space between the wall of cylinder and imaginer spherical sector (2); imaginer spherical sector (3); spherical surface (4); radiuses of imaginer spherical sector (5); b) this cylindrical lens with homogenous gradient layer.

If the gel-polymer matrix is viewed as a spherical segment, than at the periphery of matrix will be also damage the parallelity of isorefraction, but differences in this region will be decreased the curvature radius of isorefractive layers. The analogous judgment may be spread for the matrix with concave spherical surface [61].

The elucidation of morphology and stratification of isorefraction peculiarities at the process of formation of inhomogeneous gradient layer in GRIN-elements with various geometrical shapes is real. On the basis of mentioned above, it is proposed the formation of inhomogeneous gradient layers based on the above effect [61]. However because of some difficulties of considered the process, its mathematical model (in the form of differential equations) is not yet created [61].

The technology of production of "ideal" un-aberration spherical lenses with axial gradient refraction index on the basis of mathematical modeling has been researched [62]. The improvement of the optical properties was achieved by creation of inhomogeneous (with various thicknesses) layer.

The caring out of following consequential operations it is suggested:

1. Manufacture of half-infinite plate-like bottom-matrix;
2. The exchange diffusion of monomers with various molecular refractions through the flat surface matrix;
3. Completion of the copolymerization process.

On the basis of mathematical calculations for a given systems e.g. DAIP-MMA) was shown that by cutting out of spherical surface (spherical lens) with the radius ***R*** from side of flat surface of half-infinite cylinder so, that the distance from sphere center to it surface up equal to ***0.95 R***, it can be obtain the gradient lens with improved optical characteristics (e.g. with essential decreasing of chromatic aberration) as compared with homogeneous lens made from the same material and with same geometrical sizes [62].

The systematical researches on establishment of the controlling factors associated with the formation of refractive index distribution for planar structures have been carried out [60, 62]. On the basis of careful analysis of data obtained it was showed that the possibility exists of the control of the **RIGP** formation process in the planar polymer structures by change of the thermal treatment regimes, composition of matrix and monomer-diffuser (penetrant) mixture, ratio of molar volumes and polymerization initiator concentration, etc. [62].

Improved criterions were established for the estimation of the quality **of selfocs** [52, 53]. From the approaches considered in literature the choice of a suitable function for optimization of selfocs production technology is more valuable. In [52] the mean square derivation of real or theoretical diffusion curves of **RIPG** from parabolic is used as such function:

$$\sigma = \sqrt{\frac{\sum_{i=0}^{n=1}[n_0 \operatorname{sech}(\alpha r_i) - n(r_{i)})]^2}{N}}, \tag{1}$$

were n(r_i)-values of refraction index in the point of radius ***r*,** received theoretically or experimentally; ***n*** and ***n_o*** — the value of this parameter over the distance ***r*** from axis and on the axis of optical element; ***N*** _ experimental pointes number on the curve of **RIPG**.

It was shown [52, 53], that this method permits to optimize the technological conditions of gradient elements manufacture for any concrete application.

The result of this approach is the impossibility of definition the quality of selfocs under equal conditions. In particular, at one and the same values of ***σ*,** the focusing properties of selfocs may be essential different. Most correct ones consider the estimate of optical properties of selfocs by method of number calculation of the ray speed in selfoc by using Euler equation at the given RIPG:

$$n = n_0(1 - \frac{1}{2}\alpha^2 r^2), \tag{2}$$

were ***a*** is constant ; ***n*** and ***n_o*** – the value of this parameter over the distance ***r*** from axis and on the axis of optical element;

Derivation of the dependence **n=f(r^2)** from it true character at various points of ***r*** proves the deviation from the ideal focusing distribution of the refractive index (**RID**). It must be noted that to estimate of selfocs quality in practice it is usual to use the numerical aperture, the values of longitudinal spherical aberration, etc. [49].

The useful information about mutual correlation between **RID** character and optical properties of GRIN-elements give the results of investigations carried out using a Mach-Zender interferometer to study the formation of **RIGP**. The mathematical model, allowing defining an optimal regime of gradient-formation has been elaborated [52, 54].

The simple method of the refractometric analysis has been described [63, 64]. Using this method, it is possible, to carry out a quantitative description of radial distribution of refractive index. It was shown, that the dependence of refractive index of GRIN-element on its radius and molar part of monomer-diffuser has symbatic character (Figure 4) [63, 64].

The main reasons for the **RID** deviating from the real systems from an ideal-focusing distribution may arise because of some technological difficulties in the process of gradient-formation. Among these reasons it is essential the bad compatibility of the matrix monomer and penetrant; the decreased diffusion rate of monomer with low refractive index into gel-polymer matrix as compared with homopolymerization rate, results in this process forestalling the diffusion and the transfer to the glassy state occurs prematurely, without achievement of ideal-focusing profile of refractive index distribution; the violation of technological regime by exchange diffusion or consolidation of created ideal-focusing **RID** [12, 27, 52, 54].

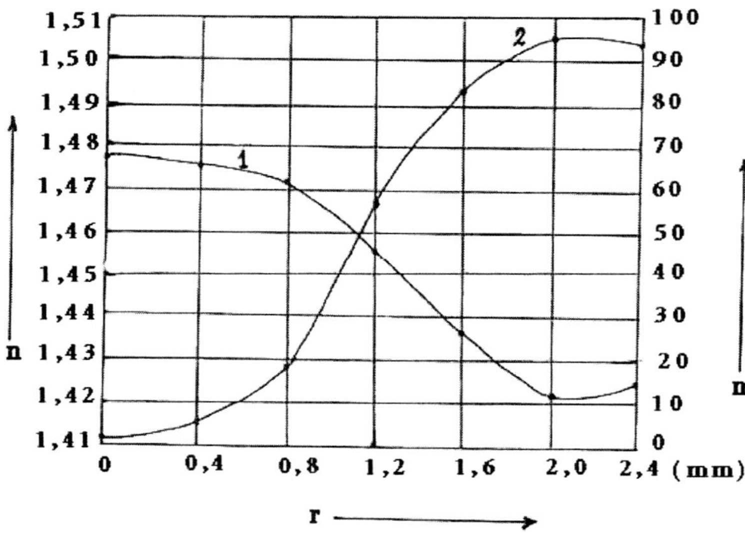

Figure 4. Dependence of value of refractive index (1) and molar part of diffusant-1,1,3-trihydrotetrafluoropropilemethacrylate (4FMA) (2) on gradient element radius on the basis of matrix of silicon-organic dimethacrylate.

Based on this data the processes of gradient creation, some authors [12, 49, 53, 65, 66, 67] have made very useful recommendations on the improvement of the diffusion technology for the production of selfocs with given optical characteristics and allowed the modernization the exist methods and preparation of a new generation of selfocs materials. From the technologies of GRIN-elements production it must be distinguish the method, founded on the closed extrusive technology [65] and so called "swollen-gel polymerization" method [66], which gave the possibility of production the GRIN-elements with excellent optical characteristics, in particular, with high value of gradient of refractive index. These elements may be used in the contemporary opto-electronics. It must be noted also that the method "crooked stencil" and the modern method of interface diffusion polymerization [67], which allows the gradient elements with large sizes [d=70mm] at $\Delta n \geq 0.02$ of special destination to be obtained, e.g. for creation of thin ophthalmology lenses without the spherical aberration and multi- focusing characteristics. These elements may be used for manufacture of optical fiber gradient elements capable of transferring optical signals with very high speed at short wavelengths [68].

The gradient elements can be obtained by elaborate of modernized methods, which control the signs of both the exchange diffusion method, and layer by layer copolymerization method [12, 39, 60, 69, 70, 71]. In these methods the liquid monomer components – gradient-formers (individual substances or blends) with various values of refractive index are used.

The gradient of concentration is created by exchange diffusion process between these monomers. It is essential to distinguish the sign – equality of the components densities. The experimental description of fabrication technology of GRIN-elements by this method has been proposed [12, 39, 69]. A metallic or glass pipes of known size, is placed on the axis on the well polished bottom of the cylindrical reactor which is flooded to a definite height with component-1 which has a high value of the refraction index, and in the space between pipe and reactor to the same height with component-2 which has equal density and a smaller value of the refractive index. After this process is completed the pipe is carefully taken out. The space occupied by the pipe is filled by the components. In the zone where the components meet mixing occurs and a cylindrical pipe is generated which contains both components 1 and 2. Inside of the pipe structure there appears a cylindrical pillar of component-1, and outside - cylindrical ring of component-2. Increasing the extent of the mixing zone width, naturally, reduces the width of the pipe wall to zero.

It is shown [17] that, the aforementioned process (at some of his experimental conditions) may be described mathematically. For example, when at filling space engaged by pipe after its withdrawal, the levels of components 1 and 2 are decreased identically, and the inner-radius (R) and the width of ring-like space between pipe and glass (r) increases on half-width of pipe wall (d), the mathematical equation of this condition has following view [17]:

$$\frac{R^2}{r(2R+4d+r)} = \frac{(R+d)^2}{(2R+3d+r)(d+r)} \tag{3}$$

At stand-by the system undergoes exchange diffusion between the components. The concentration gradient is fixed by the copolymerization reaction and the gradient of composition consequently, is the gradient of refractive index. The regime of fabrication and copolymerization was established experimentally [12, 39]. By this method GRIN media with circular cross section, from which may be formed the polymer optical fibers with gradient of refractive index, are manufactured.

The method of obtaining of GRIN media with elliptical cross section, from which one can form the one-mod light conductor (optical fiber), has been also described [12, 39]. One obtains such selfoc using analogous methods (installation) in which the pipe and reactor have the elliptical section [39].

Above examples produced selfocs with a radial gradient of refractive index. From a practical point of view, it is desirable to be able to create optical lenses with the gradient of refractive index in both at radial and axial directions [12, 60, 72]. These publications propose that such lenses can be produced use of an impermeable diffuser mask. This method is interesting in connection with the development of planar lens technology for microelectronic industry [12].

2. THE POLYMER MATERIALS WITH THE GIVEN GRADIENT OF REFRACTIVE INDEX. CARRY OUT OF THE GRADIENT-FORMATION PROCESS UNDER INFLUENCE OF VARIOUS PHYSICAL FIELDS

Diffusion copolymerization and photopolymerization are the main methods used for manufacture of GRIN-elements and allow the creation of a parabolic distribution of refractive index gradient. The common essential limitation of both the above methods is the limitation of choice of starting materials. Besides the method of creation of an inhomogeneous diffusion surface layer, including additional hard transformation, as in the case of corresponding inorganic materials based on quartz-optical glass.

The search for ways to perfect the existence technology and creation of new technological methods is a constant quest by researchers engaged in gradient materials [17].

The main principles and techniques for creation of macro-surface heterogonous gradient-formation and the methods and regimes of production of macro-surface gradient elements/media with given gradient of refractive index have been described [12, 39, 42, 54, 73-86]. It was shown, that formation of the radial RID on the basis of one and the same process, two principally different results were obtaining for polymer medium with properties of collecting and dispersing lens, produced by the introduction of gradient-former to the reaction zone by various methods [17].

The gradient formation occurs by interaction of gradient-former (e.g. diffuser with refraction index n_1) with the gradient-carrier. The function the gradient-carrier carries out is to modify the gel-polymer matrix with the refraction index $n_2 \neq n_1$. This substance must generate the optical transparent copolymer with the other monomer which has a different refractive index by a two stage copolymerization with gradient-former. The requirement of stability of the initial form and acquired properties of the gradient carrier have been considered. In the some cases the liquids, inert to the gradient-former, and the initial polymers used. The density of the liquid used is higher or lower relatively to gradient-carrier in dependence on the character of the given task [12, 74]. The time required for the process is controlled by analysis of the function $n=f(\tau)$, where n-refractive index, τ – is the duration of the gradient-formation process. The condition must be chosen such that in the function $n=f(\tau)$, the maximal value of τ does not exceed a certain limit, dictated by technical, economical or other considerations. The minimal value of τ is several minutes [12].

The initial quantitative characteristics of gradient formation are:

- The value of homopolymers refractive index;
- The value of refractive index of the processed products;
- The profile which is created initially and which may have a radial (axial) distribution of refractive index, e.g. function n=f (R), where R is the radius (length) of the sample;
- The experimentally established dependence of the change of refractive index $\boldsymbol{n}$ on the duration of the process, by means of which the function $\boldsymbol{n=f(R)}$ transformed to $\boldsymbol{\tau=f(R)}$.

On the basis of these principles it is possible to prepare polymer media with a given size and gradient of refractive index [12].

Based on the above mentioned concepts it may be concluded that, by using of this method, formation of the given distribution of refractive index is due to the duration of the gradient forming process at the surface of the gradient-carrier component [12, 73, 74].

From the practical point of view, methods of the control of gradient formation on spherical surface attracts the attention of researchers because of the spherical lenses have the defects of pictures - geometrical and chromatic aberrations. The removal of such effects is usually achieved by combination of some spherical lenses with various curvatures, of refractive index and dispersion of refractive index. For this reason, the design of modern optical devices is complicated.

Modified versions of the methods for the creation of components with radial gradients are based on diaphragmation (diaphragm-separation) of the zone of gradient-forming process [12] in the centrifugal field (Figure 5) [82], whilst taking into account the curvature of elaborate lenses. This approach is of definite interest in connection with the problem of the correction of geometrical aberration.

In this review, it will be shown that it is possible to use non-traditional approaches to solution of the problem of creation of a given radial profile of **RID** in the macro-surface polymer medium by the frontal-axial influence, based on the gradient exchanged diffusion into the gel-polymer matrix with completion of the process by copolymerization under conditions where the process is controlled by the proposed algorithm [12].

2.1. Diffusion Formation of the Given Gradient of Refractive Index on the Spherical Surface of Sample in the Centrifugal Field

Two approaches have been described for the formation of the given gradient of refractive index [12, 17, 54, 70, 71, 82-86]:

1. The gradient-formation on the spherical surface of sample with decreasing of duration of the process along radius from periphery to the center;
2. The gradient-formation on the spherical surface of sample with increasing of duration of the process along radius from periphery to the center.

The gradient-formation according to the first approach may be achieved in two ways: **a)** the permanent increasing of gradient-former volume on the periphery of revolving reactor 1, or **b)** elimination of gradient-former (filling in over reactor) from the periphery of reactor (1) through tube (6) by the supply to the reactor (1) center (tube 4) of inert liquid of lower density **(Fig. 5)**.

The liquid supplied to the rotating reactor **(Fig. 5)** under the influence of a centrifugal field is throw off to inner wall (zone) and takes the form of a cylindrical pipe with wall thickness $x=a$. The pipes bottom (the spherical layer with height K) attaches to prominent lens sample (7). As far as the gradient-former is concerned, x increases from 0 to R, and K increases from 0 to H. It is necessary to determine the dependence of the gradient-former volume on x to determine the value of K. It may be shown, than when $x=a$, $K(\tau)=-(r-H)\pm\sqrt{r^2-[R-x(\tau)^2}$ (4), where τ is the duration of the process.

This equation has in a physical sense only positive value of the root. According to (4), the radial distribution with decreasing process duration on the samples surface from periphery to the center is expressed by the equation:

$$V_{gr}(\tau) = \frac{1}{6}\pi\{6hx(\tau)[2R - x(\tau)] \pm K(\tau)[6hx(\tau) - 3x^2(\tau) + K^2(\tau)], \quad (5)$$

where $\boldsymbol{V_{gr}(\tau)}$ is the volume of gradient-former at the moment $\boldsymbol{\tau}$; $\mathbf{x(\tau)}$ is the reverse function of $\boldsymbol{\tau(x)}$. For convex lenses, the second part of the equation is negative, for concave it is positive [73, 74].

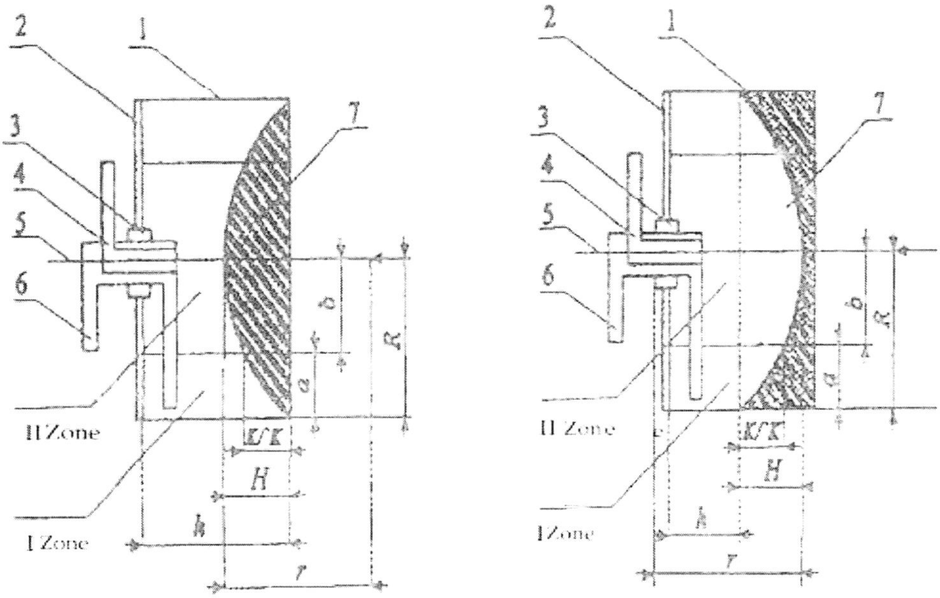

Figure 5. Main features of the device for formation of radial gradient of process duration in the centrifugal field (chemical reaction/exchange diffusion) on the lens surface (diametrical cross-section): 1 cylinder reactor; 2 lid; 3 gasket; 4 immovable pipe/air eliminator for injection/elimination of active/inert liquids; 5 reactor rotation axis; 6 immovable pipe for injection/elimination of active/inert liquids; 7 lens sample (A convex, B concave); a cylinder pipe thickness of active liquid; b radius of the active liquid cylinder; h height of the internal space of the reactor (1) cylinder on periphery; K height of the spherical base layer of the cylinder pipe (x = a) of the active liquid; K' height of the spherical layer base of the cylinder pipe (x = a) of the inert liquid; H height of the spherical segment; R radius of the lens sample; r radius of the sphere, to which the lens sample surface corresponds.

According to the second approach, in which the reactor (Figure 5), fulfills the gradient-former component, through a pipe (4), and the chemically inert liquid has less density. It has a cylindrical form in II zone, the diameter of which increased from $\boldsymbol{b}$ to $\boldsymbol{R}$. The cylinder bottom represents the spherical segment. For an inert liquid, the gradient- former component achieves the cylindrical pipe form with a spherical layer with high of $\boldsymbol{K}$. The thickness of cylindrical pipe wall $\boldsymbol{x=a}$, is decreased from $\boldsymbol{R}$ to $\boldsymbol{0}$, and $\boldsymbol{K}$ is decreased from $\boldsymbol{H}$ to $\boldsymbol{0}$. According to (4) the decreasing radial distribution process is expressed by equation:

$$V_{inert}(\tau)=\frac{1}{6}\pi\{3[R-x(\tau)]^2[2h\pm K(\tau)\pm H\}\pm\{H-K(\tau)]^3\},\qquad(6)$$

where $V_{inert}(\tau)$ – the volume of inert liquid at moment τ. The minus/plus signs correspond to the prominent/concave lens [73, 74].

The gradient-formation, according to the second approach, can be achieved in two ways. The first approach is to firstly fully fill the reactor with the gradient-former, which after is eliminated from center of reactor (pipe 4) by an inert liquid which enters by pipe 6 and has a larger density. At this moment the gradient-former component takes the form of a cylindrical pipe with the shape of a spherical segment with height of ***H***. The cylinder radius of gradient-former component (liquid) ***x=b*** is decreased from ***R*** to ***0***, and height of the cylindrical pipe of spherical layer ***K'*** is increased from ***0*** to ***H.***

It can be shown, that when ***x=b,*** $K'(\tau)=-(r-H)\pm\sqrt{r^2-x^2(\tau)}$ **(7).** In this case the physical sense has a positive value of the root. The radial distribution of process duration is expressed by equation:

$$V_{inert}(\tau)=\frac{1}{6}\pi\{[6h\pm 3K'(\tau)][R^2-x^2(\tau)]\pm K'^3(\tau)\}\qquad(8)$$

According to second method, the gradient- former with the lower density is provided through pipe (4). In zone II, the radius of the cylindrical gradient-former ***x=b*** increases from ***0*** up to ***R*** and the height of spherical layer ***K'*** of the foundation of the cylindrical pipe of the chemically inert liquid (I zone) decreased from ***H*** to ***0***. The inert liquid is eliminated from periphery of the reactor through pipe 6. The radial distribution of duration of the process decreased by equation:

$$Vgr(\tau)=\frac{1}{6}\pi\{3x^2(\tau)[2h\pm K'(\tau)\pm H]\pm[H-K'(\tau)]^3\}\qquad(9)$$

The minus/plus signs correspond to the convex/concave lens.

The analogous approach can be used for the creation of flat parallel lens with given radial **RID** [12, 82].

It must be noted that the control of the gradient-formation process on the spherical surface can be also be carried out in a magnetic field, when the chemically inert liquid is used.

2.2. The Forced Diffusion Under Influence of Electrostatic Field with Given Inhomogeneity and Configuration

The natural process of exchanged diffusion of the monomer-diffuser into the gel-polymer matrix with space-bonded structure was considered above. In [43, 44], the method of forced diffusion of monomer with a given refractive index and with a definite dipole moment to the matrix from solid linear polymer, under the influence of an electrostatic field with a given inhomogeneity and desirable configuration, has been described. It was shown, that at co-incidence of movement of polar monomer molecules under influence of electrostatic field with the direction of diffusion, the rate of forced and natural diffusion add and the diffusion process is intensified. The coefficient of diffusion here depends not only on the temperature and viscosity of the material in which the diffusion of the polar liquid proceeds but on the value of its dipole moment and on the voltage of the electric field [43]. The electrostatic inhomogeneous fields can be created by use of electrodes with various forms. Therefore by variation of the direction of diffusion flows of polar molecules and through guidance of the configuration of the electrostatic inhomogeneous field and the structure of the diffuser, it is possible to obtained optical gradient element and devices with difficult (complex) configuration [43]. On the basis of the experimental data it was shown, that change of the field voltage on the process of diffusion, allows control of the conditions of diffusion, consequently, the value and character of the variation of the refractive index in the gradient element constructed [43]. It was supposed the principal factor controlling diffusion of the polar molecules is the strength of the electronic field. The distribution of the diffuser molecules in the polymer cylinder can be described by $C(t_1,r) \approx const \bullet r^2$ (r – the distance from center to point being considered). By using of the Ber-Gladston-Dail formula for the refractive index ($n=1+const \bullet N$, where N is the "density" of the investigated substance particles) one obtains the dependence of the refractive index of the polymer cylinder on the radius R: a) before diffusion $n_0=1+(const)_0 \bullet N$ и б) for the diffuser $n_1=1+(const)_1 C(r,t)$. By using of the Principle of Additivity of Landolt refraction, one obtains the equation for refractive index of a cylindrical bar with dipole-phorez of polar molecule enriched to the center [43]:

$$n(r,t) \approx n_0+(n_1- n_0)\zeta(t_1)r^2, , \tag{10}$$

where $\zeta(t_1)=\dfrac{C_0}{NR^2}\sqrt{\varphi(t,r)}$.

This equation shows the growth of the refraction index of the polymer cylinder after dipolephorez and decreases towards the center depending on the relative values of the refraction index of the matrix and diffuser in proportion of the squared radius [43].

The concrete method of obtaining the gradient element with given RID based on styrene-MMA, by using of the dipolephorez method, has been described [43]. On the basis of interference investigations (using a Max-Zender interferometer) it was shown, that the penetration deep of the diffuser to the cylindrical polymer, in the inhomogeneous electrostatic field (cylindrical condenser) with a given inhomogeneity and configuration is about three time larger, than in the sample without the influence of the electrostatic field [43].

This method is essentially different from the one used for flat inorganic micro-lenses with gradients of refractive index in the glass based on electro-diffusion of ions, which changes the gradient index of the original material (inorganic glass) [43].

The above approach may be used in the case of the application of a magnetic field to force the diffusion of the substance with a magnetic moment into the solid polymer material [43].

The authors [43] noted that in the case of changes in the values of the properties, besides the optical ones, for a given area of medium under the influence of an inhomogeneous electric and magnetic fields, it is possibility to create the diffusion flows leading to change of coefficients of conductivity, heat conduction, viscosity, friction, elastic module, etc. This method can be used to manufacture of materials with given properties based on industrial polymers [43].

3. Nontraditional Methods of Gradient Formation on the Basis of Solid Materials

In the previous section, some solutions to obtaining polymer gradient elements (media) were discussed. Most of the methods are based on exchange diffusion of penetrant and residual monomer of gel-polymer matrix in the direction of refractive index gradient coinciding with the direction of diffusion, which is essentially limited by the diameter of the gradient elements. The axial influence of the penetrant and use of diffusion non-penetrated plate in the element, simultaneously formatted the gradient both in radial, and in axial directions [12], which completes the perfect analytic expression describing the ideal-focusing distribution of refractive index in such real systems. In special conditions, there is the possibility to successful overcome these limitations, characterized by the exchange diffusion and produces GRIN-elements with a large diameter (≥70 mm) both for radial and axial influences [17].

3.1. The Gamma-Radiation Method of Gradient-Formation

The purpose of this section is to show the possibility of creation of a macro-surface gradient media with given RID by influence of gamma-radiation on some optically transparent industrial polymers (by means of investigation of post radiation change of mutual bonding values-absorption ability in ultraviolet region of spectrum and refractive index in the visible region of spectrum) [12, 28, 75, 76]. For this investigation, the polymers were chosen which take into account their transformations under the influence of ionizing radiation: destruction of branched main chains or net structure, these polymers are: polymethylmethacrylate (PMMA); polyvinylalcohol (PVA); isotactic polypropylene (PP)], for which radiation degradation is compensated by "scission" of the main chains. The source of gamma-radiation ^{137}Cs provides irradiation at a level of 2.5 Gr/min, the doze of irradiation R≤60 kGr. The irradiation was carried out in air at 25±3^{0}C.

The ultraviolet absorption spectra are given (fig 6) for the initial and irradiated by dozes R=28.8 kGr of polymers' samples. All three polymers have the strong absorption in the spectral range (59-45)10^3cm^{-1} (190-220 nm), the tail of which is spread till $\tilde{\upsilon}$=27•10^3cm^{-1} (λ≈370 nm) and in the visible range they are all transparent after irradiation. At such dozes of irradiation in the range of R≤60 kGr for all these polymers, it was observed increasing of the absorption in spectral range 190÷220 nm [76].

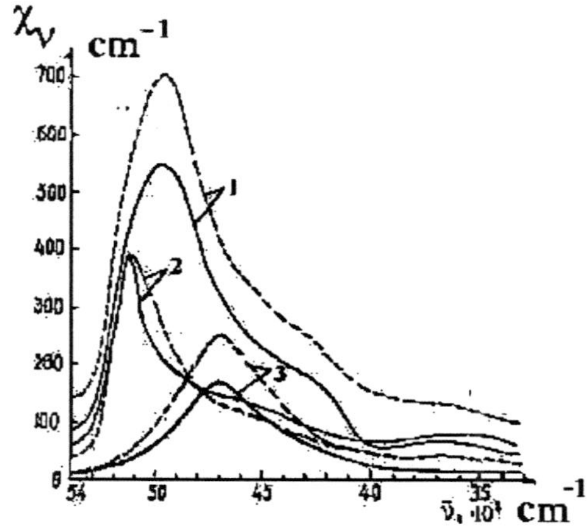

Figure 6. The absorption spectra of in ultraviolet region for the initial and irradiated polymers: 1- PP; 2-PVA; 3-PMMA (The solid line –initial samples; the dashed line- gamma-irradiated samples; æ$_\upsilon$- common losses)

The fixed wave lengths in figure 7, correspond to (1) and (3) to the maximum of the absorption curve, (2) - to the relevant maximum, which is coincided with peak of the absorption curve. The character of the response in the view of additional absorption of all three polymers is nearly identical. If the ratio of the additional absorption $\Delta æ_\upsilon$ to the main zone of absorption will be a degree of steady, than this ratio may be present so: $(\Delta\chi_{PVA\,(1)})^{-1} : (\Delta\chi_{PMMA(1)})^{-1} : (\Delta\chi_{PP(1)})^{-1} = 30:3:1$.

The gamma-irradiation of PMMA is accompanied by the appearance in the polymer structure of double bounds of the vinylidene type, which are obtained as a result of homolytic rupture of the main chains of this polymer and consequence transformation of the generated intermediate particles [17]. In the initial stage of gamma irradiation, up to dozes of irradiation ***R≤7.2 kGr***, the process of the accumulation of vinylidene type double bonds takes place.

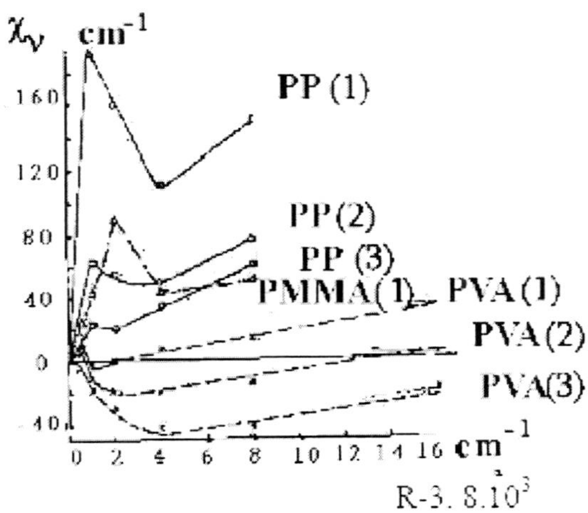

Figure 7. The dependence of additional absorption on the irradiation doze for three fixed wave lengths of maximums: PP (1) - 204 nm; PP (2) - 228 nm; PP(3)- 278nm; PVA (1) -195 nm; PVA (2) - 217 nm; PVA (3) - 282 nm; PMMA(1) – 212,8 nm.

The absorption of **PVA** in spectral range **190-300 nm** is due to carbonyl chromophoric groups, which are obtained as a result of oxidation process with participation of atmospheric oxygen [76]. The small increase in transmission in the spectral range of **185-193 nm**, as indicated by the dependence of $\Delta æ_{\upsilon}=f(R, \upsilon)$ is in the case of **PMMA** due to the dynamic and concurrent processes of destruction and scission of the polymer macro-chains under the influence of gamma-irradiation ($æ_{\upsilon}$ – the spectral index of common loses of absorption and scattering for given sample) [76].

In the **PP** the loss in light intensity $æ_{\upsilon} = f(\upsilon)$ at frequencies $\upsilon=49.8\cdot10^{3}cm^{3}$ **($\lambda_{max}=204$ nm)** in the main absorption is defined by both absorption and contributions to the process from scattering from crystalline regions in the sample. The accumulation rate of vinylidene groups from destruction of the polymer macro-chains is higher for low dozes of irradiation. For all three polymers, the vibrations correspond to intermediate structures and the completing processes of accumulation and consumption [76]. These vibrations reach their largest amplitudes in the case of PP, therefore the creation of GRIN-elements by the method of gamma-irradiation has application of this polymer.

The appearance of additional absorption in the far ultraviolet range of the spectrum by irradiation of polymers, is connected with rupture of the C–C bonds and the creation of double bonds of vinyldene type, indicate the possibility of changes in the refractive index in the visible range of spectrum as a result of the production of an additional absorption feature associated with the un-saturation. This situation is observed for **PP,** it being the most sensitive material to gamma-irradiation but has the ability to retain satisfactory mechanical properties [75, 76]. For the **PP** films, it was established that post-radiation change of refractive index occur. It was shown, (Figure 8) that the value of refractive index of irradiated **PP** increases with increasing level of gamma-irradiation doze and at $R\geq14.4$ ***kGr*** reached saturation with $\Delta n \approx 0.017$ (wave length $\lambda_{max}=632,8$ ***nm)*** [76].

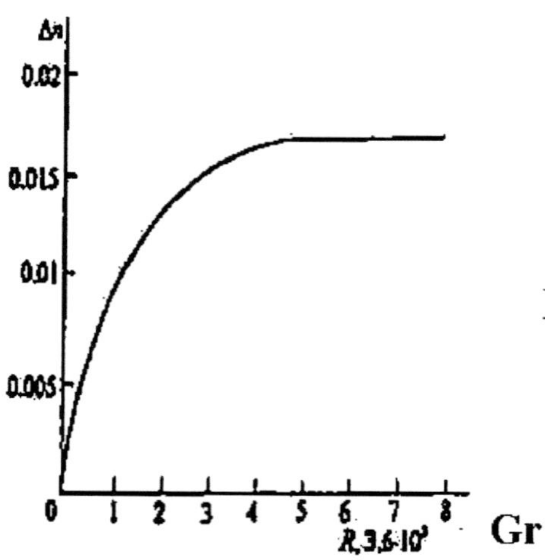

Figure 8. Change of PP film refractive index under of the various doze of gamma-irradiation

The creation of any **RID** profile in the gradient medium corresponds to a distribution of the gamma-irradiation doze on the surface of the sample of polymer and is dependent on three factors [17, 76]:

1. The experimentally established dependence $\Delta n = f(R) = \Delta n_{max} R / R_{max}$ (Figure 8);
2. The dependence on saturation of the irradiation power with decreasing (k) on the thickness of mask (d) and the calculated profile (which is placed between gamma-radiation source and PP film), that is $k= f(d)= e^{æd}$ (k = R_O/R, where R_0 и R – respectively the dozes of gamma-irradiation, falling and passing though the lead mask), which is defined for the energy of irradiation quanta of ^{137}*Cs* ***E(hυ)=0,662 Mev*** by the literature date [76]. The value of the absorption coefficient is **æ≈0,087 nm^1*;***
3. For a given dependence of ***DRI n=f(R),*** where ***r*** is the length/radius for sample [76]. The gradient in the medium with given ***DRI*** is achieved by displacement of lead mask with the defined profile d=f(r) between the gamma-irradiation and polypropylene film, at $n(R)=n_o + \Delta n_{max} \bullet e^{æd} = \frac{R_{\max}}{R} \approx 100$, which indicates that **æd=ln100**, d=53mm [76].

This approach allows one to create by gamma-irradiation of polymer films with a profile of refractive index distribution, which is a necessary condition for the creation of a profile which achieves correction of residual aberration in optical system, and the creation of the integral optics elements, light-guides and transformers of light beams [76].

This method has been tested on thin films of isotactic polypropylene and may be transferred to other type of radiation-stable optical transparent polymers, where it is possible to change the refractive index at definite doze of gamma-irradiation and their large application in the optical engineering [17, 76].

3.2. The Disk-Like Elements with the Structural Inhomogeneity Based on Powder Materials

The using of powder materials for the creation of gradient systems may be proposed both in the basis of some optical polymers and inorganic glasses and for other materials with the aim of management of physical and chemical properties depending on radial/axial profile of the distribution of materials [12, 17, 87].

Rotating around the axis of a cylindrical vessel using a defined program [88] and a blend of powder-like substances can produce gradient materials. After melting of the components and the formation of monolith, the material is allowed to settle. The settling process is increased from the periphery to the center and is proportional to a decrease in the density (ρ) of the material filling the vessel. The settling process for a structure with height *H* and with the exception of horizontal transference of melted mass, the gradient is increased from the periphery to the center along the radius. The following equation has been used [12]:

$$\Delta Hj = 4(2j+1)\{[\frac{\rho_1}{\rho_1(R)} - 1][H - f(j\Delta r + \frac{\Delta r}{2})] + [\frac{\rho_2}{\rho_2(R)} - 1] \bullet f(j\Delta r + \frac{\Delta r}{2})\}, \qquad (11)$$

where ΔHj – is the increase of height of the cover of the cones-like vessel on the j-interval; H - the height of cylindrical storage; Δr – interval of permanent length of radius R; j – number of Δr – interval of radius of storage and Δr=1,2,3…; ρ_1 – the density of first component of in the monolith state; $\rho_1(R)$ – the density of first powder component in the forming vessel, depending on R radius of vessel; ρ_2 – the density of second powder component in forming vessel depending on R radius of vessel; $\rho_2(R)$ – the density of second powder component in the forming vessel, depending on R radius of vessel;

By mean of this equation it is possibility to describe the profile of the conic vessel, which leads to settling of the powder components to produce gradient materials [12, 17].

The proposed method can be used to obtain materials with a radical gradient both in refractive index and other physical and chemical parameters [17].

3.3. Polymer Materials with the Given Gradient of Birefringence

The term "gradient optics" used up to recently to describe materials in which the gradient in the refractive index is in one direction. Simultaneously, in the 1990's some researchers [17, 88-90] expressed the opinion that the gradient of other optical properties (e.g. dispersion and birefringence) can be used in the optical device-building industry. The materials with an axial birefringence were described for the first time by [17, 88, 89]. They were named as GBR (gradient of birefringence) materials. Introduction to the optics of new parameter, the gradient of the birefringence changes principally the sense of gradient optics and expands the sphere of it investigation. Now the "gradient optics" includes two directions: a) the optics of selfocs [GRIN (gradient refractive index) optics] and GB (gradient of birefringence)-optics [12].

The polymer structure as a result of a deformation gradient becomes anisotropic and has the property of birefringence. A series of devices for production of GB-elements have been described [12, 17]. These devices were created with the application of an inhomogeneous mechanical field to create inhomogeneities, which have the effect on the width of the polymer films and create a gradient of orientation in the structure (and, consequently, a gradient in the birefringence) [89]. Some designs are described in (Figure 9). They are distinguished one from other by the characteristics of the applied mechanical field.

On the Fig. 9,a the left clamp installation, for which $\alpha_{max}=180^0$ (the second clamp here and on the following figures presents the mirror representation of the first clamp) is shown.

On switching on the electrical motor (1) causes the teeth-like wheel (2) to turn (which is connected with the bar 7 by means of bar 5) on $\boldsymbol{\alpha \leq 180^0}$ stretches the material in the indicated direction and creates a strain by stretching the material, the spherical edges (4, 9) of the clamps (6) in the nests (3, 8) will be turned in such a manner, that the edges of clamps will be turned relative to one to another by the process of stretching. Such functions are carried by the spring (10), connecting the lower edges of clamps. Therefore the slipping of sample (11) on the clamps is excluded. In this installation the stretching deformation of sample (11) as it is in Fig. **9,b**) increases from above to below.

The scheme for another installations is shown on Fig. **9,d.** This installation uses a rubber clamp (16), in which the film being studied (2) is placed. The clamps edges (16) are fixed in the frame (12). The pattern (4) is fixed on the frame (12). A shaft (11) on which the system rests; consist of a wheel (8) with channels and ferrado (9). A rope (7) is fixed to one edge by a clamp (3) and to a second by a wheel (8). On switching on electrical motor (17), the rope winds on the wheel, to which the rubber clamp is attached and advances the stencil step by step. When the clamp reaches the stencil the stretching of the film is finished, but the ferrado continues rotating. By choice of stencil configuration the desirable distribution of deformation can be achieved in the films.

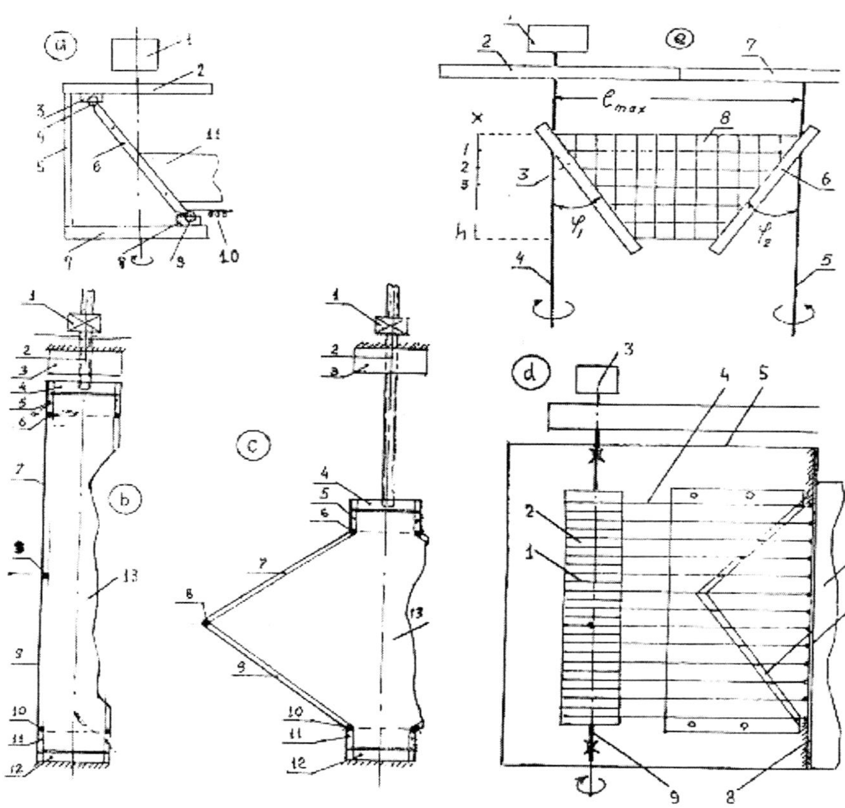

Figure 9. Devices for gradient orientation of polymer films (description see in the text).

One of the models shown in Fig. **9,e** involves the sample of the polymer (film, plate) which has a trapezium structure being placed between the two clamps (3, 6) with non-parallel squint edges. Demonstration of the character of the lengthening distribution it is visible by laying a net on the polymer sample. The clamps are rotated by mean of an electric motor with gearing (1) and a teeth-like wheels (2, 7) the clamps (3, 6) rotate in the contrary directions around the parallel axis (4, 5). At rotation of the clamps (3, 6) causes their edges to be displaced through the side surface of the cones, which leads to a distribution of lengths in the width of the sample (increases from above to below) and, consequently, the GB-effect is created [89]. In the described installation, the maximal turning of the clamps (3, 6) $\boldsymbol{\alpha_{max} = 90^0}$. At $\boldsymbol{\alpha > 90^0}$ the sample of polymer film will lay along the clamps edges.

It can be concluded, that various approach exist which can be used to control the inhomogeneity of the mechanical field: in various equipment (Fig. **9,a** and Fig. **9,e**) rotation of clamps between $\boldsymbol{0 \leq \alpha \leq 90^0}$ produces the desired field. Simultaneously two types of deformation-longitudinal elongation and cross-sectional compression (Fig. **9,b**) can also be used. In the device shown in Fig. **9,d** the deformation is one of longitudinal elongation. The additional variation of the mechanical field inhomogeneity is possible by the use of clamps with any configuration [12].

The properties of the GB-elements (materials) are illustrated below on the samples deformed in the arrangement shown on Fig. **9,e**. The samples form shows that the change of distance between clamps edges (3, 6) (Δl) exactly correspond to the form and sizes of sample, the increasing of length and width (Fig. 9,e) in process elongation at φ_1 and φ_2 is described by following equations:

$$\Delta l = 2x \bullet tg\varphi(1 - \cos d) \quad (12)$$

and

$$\Delta h = h\sqrt{1 + \sin^2 \alpha \bullet tg^2 \varphi} - 1) \quad (13)$$

It is important to estimate the conditions of for the inequality ***Δl/2>Δh***, as the creation of the controlling principal for longitudinal orientation of ***n*** and, consequently, the application of this approach to produce materials. It can be shown, that this inequality is equivalent to an inequality: ***cosα<1-Δl/2R***, where ***2R*** is the radius of bottom of imaginary cone, on surface of which displacement of the clamps occurs (3, 6) and Δ***h***< ***R***.

The elongation uses the complex connection between control of the inhomogeneity of the mechanical field and the topographical picture of the elongation deformation distribution. The region of usable deformation, which is localized in the center of the sample gradually expands and takes up the whole width of the sample.

The investigation of the polarization characteristics of the gradient-oriented polymer films was carried out by means of a ***Berek*** compensator and a polarization microscope ***MIN-B*** (wavelength equals to 0,636μm (636nm), using a special fabrication laboratory [90].

In Fig. 10, the curve expressed the distribution of the elongation on the width ***h*** of a **PVA**-film. The numerical values were taken for the most deformed region of the film. The difference of the elongation is about one order of magnitude**.** In the arrangements presented in Fig. **9,b** and **9,d** can increase the difference by 2-3 orders of magnitude or more. On the same figure, the curve 2 expresses the dependence of the thickness ***d*** on the width ***h*** of the sample. The values are taken on the zero line placed at the central part of the sample. As the elongation is increased, the thickness of sample is decreased.

The curve 3, on Fig. 10, expresses the distribution of GB on same line of the sample width (***h***). The axial gradient of GB is clearly expressed in relation to the characteristics of lengthening or the orientation and thickness of the film (curve 2) [90]. The polarization properties of the GB-elements were investigated [91]. Investigation of the polarization characteristics were carried out using the method described in [92]. The initial materials for the GB-element formation, was PVA film with the thickness equal to 100-150 μm. Heterogeneous mechanical tensioning of PVA film samples in an equilateral trapezium form with the bases l_{max} = 72mm and l_0 =10mm and height **z**=33mm along the base direction was carried out [93].

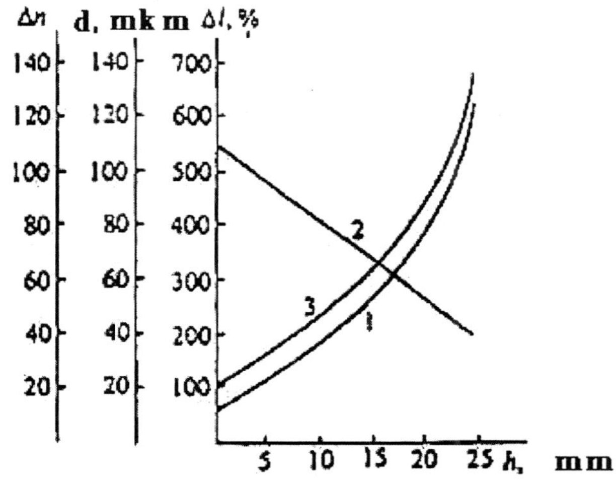

Figure 10. The distribution of lengthening (1), thickness (2) and gradient of birefringence (3) along width (h) in GB-element based on PVA.

The tension temperature of the polymer films was 80-90^{0}C and the angular velocity of the clamps rotation was equal to w=0.0785 sec^{-1}. Automatic registration of the intensity of light passing through the film using a He-Ne Laser light-beam (λ=632,5nm) was scanning along the direction **z**, which in the center of the GB-element coincide with birefringence direction (Laser light-beam polarization plane angle equal to 45^0). A record of the intensity was made across the film {$\mathbf{I\perp = I_0 \bullet sin^2 [(\pi/s)\ \Delta n \bullet d]}$} and parallel {$\mathbf{I_{II} = I_0 \bullet cos^2 [(\pi/s)\ \Delta n \bullet d]}$ (**Δn**-gradient of birefringence, **d**-the thickness of the polymer film**)** to the polarization plane of the Laser beam and analyzer [91].

The relation between the intensities **I⊥** and $\mathbf{I_{II}}$ can be expressed by a formula $I\perp/I_{II}=tg^2(\pi/\lambda \bullet \Delta n \bullet d)$. Using this equation and the experimental data I⊥ (*z*) and I_{II} (*z*) and the measured thickness [d=f(*z*)] of the polymer film sample, the gradient of birefringence Δ*n,* along the scanning direction for any point of GB-element, was calculated [91].

The diagram of ***Δn=f(z)*** are show in Figure (11), for the central part of the GB-elements based on **PVA** with an initial thickness d_o=110 μm. This GB-elements was made at 80^{0}C with different degrees of the maximum relative lengthening $\eta_{max}=\Delta l_{max}/l$

At η_{max}=1,1-1,6, the relation ***Δn=f(z)*** includes two linear section – ***a*** and ***b*,** with a grad(Δn) = d(Δn)/dz=tgθ and $tg\theta_a = 0.07 \bullet 10^{-3}$ mm^{-1}, $tg\theta_b = 0.6 \bullet 10^{-3}$ mm^{-1}[91]. It is possible to choose the third liner section with $tg\theta_c = 0.95 \bullet 10^{-3}$ mm^{-1}at η_{max}=3.35-3.4 and the forth one with $tg\theta_e$ = 1.7 at •10^{-3} mm^{-1}at η_{max}=6.8-7.25. The maximal linear velocity of the zone with Δn gradient formation was changing and was equal to $V_{1(max)}$=2.5 mm /sec, $V_{2(max)}$=2.2 mm /sec, $V_{3(max)}$=1.9 mm /sec when z_{max} was decreasing in dependence on η_{max} : z_{max}= 32 mm, z_{max}= 28 mm, z_{max}= 24 mm. The observing deviations from the linearity in ***a, b, c, e*** sections can be associated with molecular structure orientation processes [91].

The more complicated character of the change of the birefringence ***Δn=f(z±δ),*** Figure (12) is observed along the parallel lines (straights) form the central axis *z* (it coincides with the scanning direction and birefringence axis), which are on *δ*=11mm and *δ*=14mm distance from it. The curve 1 on Figure (12) corresponds to ***Δn=f(z)*** relation.

$$\sim CF_2-CF_2O^{\bullet} \rightarrow \sim C^{\bullet}F_2 + CF_2O \xrightarrow{NO} \sim CF_2NO \tag{54}$$

The adjacent terminal double bonds and terminal nitroso compounds formed can enter into a reaction to synthesise nitroxyl radicals:

$$\sim CF_2N{=}O + CF_2{=}CF-CF_2\sim + NO \rightarrow \sim CF_2-N(O^{\bullet})CF_2-CF(NO)-CF_2\sim \tag{55}$$

The advantage of the suggested method for the preparation of spin-labelled polymer is that the nitroxyl free-radical fragment is incorporated in the basic macromolecular chain without disturbing its orientation.

An analogous investigation of the action of NO on γ-irradiated tetrafluoroethylene-hexafluoropropylene copolymer (TFE-HFP) containing 13 mol% of HFP units has been performed [33]. After exposure of powders and films of TFE-HFP to a dose of 10^5 Gy in air, there are three types of stable peroxide macroradicals:

1. End radicals $\sim CF_2\text{-}CF_2O_2^{\bullet}$ (denoted $R_eO_2^{\bullet}$);
2. Secondary mid-chain radicals $\sim CF_2\text{-}CF(OO^{\bullet})\text{-}CF_2\sim$ (denoted $R_cO_2^{\bullet}$);
3. Tertiary mid-chain radicals $\sim CF_2\text{-}C(CF_3)(OO^{\bullet})\text{-}CF_2\sim$ (denoted $R_tO_2^{\bullet}$).

Their total concentration is $[RO_2^{\bullet}] = 3 \times 10^{-3}$ mol/kg, of which (25 ± 5)% are tertiary peroxy radicals. Under the action of NO on evacuated samples, the radicals decay to form peroxy radical conversion products and tertiary nitroso compounds:

$$\sim CF_2-C(CF_3)(NO)-CF_2\sim$$

Heating these samples in vacuum up to 473 K leads to the formation of nitroxyl radicals of the type:

$$\sim CF_2-N(O^{\bullet})-CF_2\sim$$

The nitroxyl radicals appear in the temperature range where the tertiary nitroso compounds decay in vacuum with the generation of tertiary alkyl radicals ($R_t^{\bullet}$). The first step of $R_t^{\bullet}$ formation is β-scission by the reaction:

$$\sim CF_2-C^{\bullet}(CF_3)-CF_2-CF_2\sim \rightarrow \sim CF_2-C(CF_3){=}CF_2 + {}^{\bullet}CF_2-CF_2\sim \tag{56}$$

In the presence of NO formed upon decomposition of the tertiary nitroso compounds, the terminal alkyl radicals can be converted into terminal nitroso compounds, which react with the adjacent double bonds to form nitroxyl macroradicals:

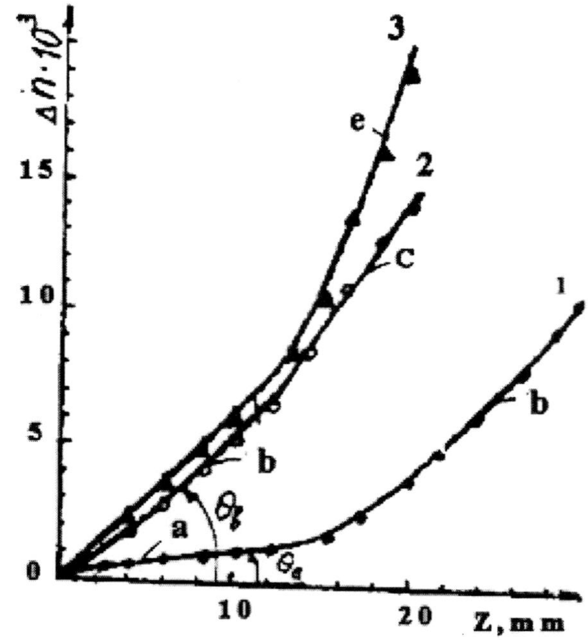

Figure 11. Dependence Δn=f(z) for different η_{max}=Δl $_{max}$ /l.: 1.-η_{max}=1,1-1,6; 2.- η_{max}=3.35-3,4; 3.- η_{max}=3.35-3,4. $tg\theta_a = 0.07 \cdot 10^{-3}$ mm^{-1}; $tg\theta_b = 0.6 \cdot 10^{-3}$ mm^{-1}; $tg\theta_c$ = 0.95 at $\cdot 10^{-3}$ mm^{-1}; $tg\theta_e$ = 1.7 at $\cdot 10^{-3}$ mm^{-1}; t=80^{0}C.

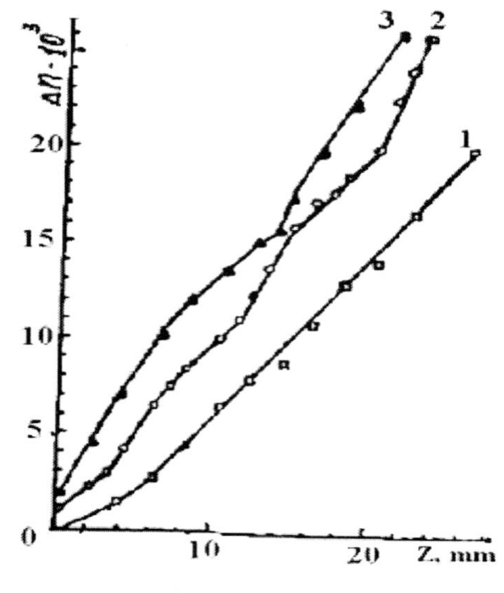

Figure 12. Dependence Δn=f(z±δ) for different δ: 1.- δ=0; 2,- δ=11mm; 3.-δ=14mm; t_2=90^{0}C; $tg\theta_1 = 0.8 \cdot 10^{-3}$ mm^{-1}.

The relationship changes spasmodically with the inclination of the *tgθ* to the birefringence axis *z* (curves 2 and 3, figure 12). It exceeds ***tgθ = 0.8•10⁻³ mm⁻¹*** for *Δn=f(z)* in the central GB-element obtaining at 90^{0}C [91]. The character of the 2 and 3 curves (Figure

12) is forming the peripheral zone of this element because this increases the deformation in the *z*-axis perpendicular direction.

The dependence of $\Delta n=f(z)$ of the GB-element on the β-angle between of Laser light-beam and plane (flatness) of this element was investigated [91]. As one can see from Figure 13, a,b,c, d, when the angle β increases, the character of the $\Delta n=f(z)$ relation does not change, but $tg\theta$ decreases: $tg\theta_a = 0.85 \cdot 10^{-3}\ mm^{-1} > tg\theta_b > tg\theta_c > tg\theta_e$. It is associated with the increasing of the effective thickness of the GB-element sample at the constant value of degree of orientation. In Figure (13), it is easily seen that the character of the change of the effective value of birefringence relatively with β-angle for different values of the *z* coordinate.

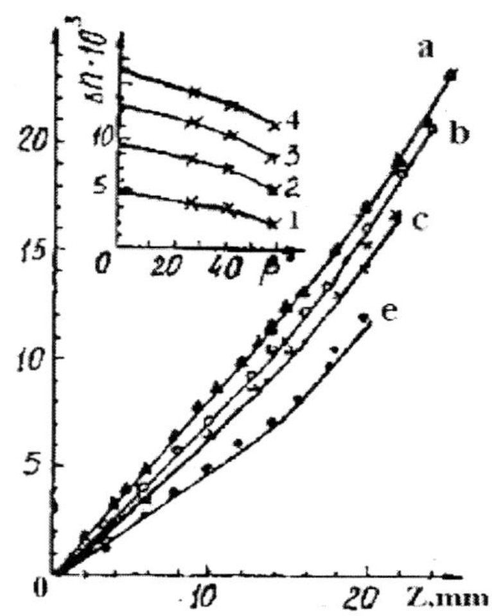

Figure 13. Dependence Δn=f(Z) from angle β between of Laser light-beam and plane (flatness) of GB-element ($\beta=0^0$, $tg\theta_a=0.85 \cdot 10^{-3}\ mm^{-1}$; $\beta=30^0$, $\beta=45^0$, $\beta=60^0$): 1.-z=5mm; 2.-z=10.5mm; 3.- z15mm; 4.-z=18mm' $t=90^0C$

As one can see from the Figures 12 and 13, by increasing the temperature from 80^0C to 90^0C used in the formation of the GB-elements, the character of the $\Delta n=f(z)$ relation becomes more linear and $tg\theta_a = tg\theta_b = tg\theta_c$ [91]. At the same time, the linearity of this relation becomes worse in the peripheral zone of the GB-elements, because of the value of the deformation in the *z*-axis perpendicular direction. The increasing of the β-angle between of Laser light-beam and plane (flatness) of this element simply leads to a decrease in the effective birefringence, but the character of the relationship does not change [91].

4. Real Application of Polymer Materials with a Gradient of Optical Properties[1]

Polymer materials with a gradient of optical properties can be the basis for the creation of optical elements with new functional possibilities. These elements will influence the future development of opto-electronics, applied and integral optics and the creation of optical

[1] This chapter was prepared in cooperation with Dr. Z. Wardosanidze (Physics).

gradient materials with given properties and possibilities of the application these materials in optical device production (Optical device building industry) for the new generation of devices. The practical application of these materials will provide solutions not possible with traditional optics based on the available inorganic glasses: large gradients in the refractive index (Δn can be 0.1 and more, when for the quartz glasses $\Delta n \approx 0.005$); low density, flexibility, low energetic capacity of process of polymer synthesis; lightness of contact and technological design of devices, etc.

The GRIN-materials with given radial/axial distribution of refractive index will be widely used in the fiber-optical communications (the gradient fibers, concordance elements), in copy machines as the light-focusing elements for Laser-systems of video-recording systems, in view of flat lenses (gradient flat-parallel lenses) and gradient thin films for correction of aberrations, in integral optical systems, etc. This will require research to be carried out on the investigation on new areas of application by creation of novel optical elements and arrangements.

At this time, in the opinion of well-known experts of gradient optics, the creation of gradient of birefringence materials is producing materials with potential application in the optical instrument manufacturing industry [93]. For the first time, the possibility of creation of optical polymer media with a given axial gradient of birefringence, using a mechanical field with controlled inhomogeneity exits [17]. At the present time, GB-element films with thickness 50-150 μm have been created, in which the values of the gradient of birefringence can be varied within the limits $\Delta n = 0 \div 0.02$, and lengthways coordinate of the gradient Δn, $z \geq 10$mm. These elements will have real application in applied optics, particularly, in the sphere of polarization optics, in polarizing optical microscopy, in photo-elastic analysis, etc.

Generally, polarimetry requires the establishment of the type, quality, azimuth and ellipticity of the polarization. In addition, polarization of some type of irradiated or dispersed light [irradiated by natural and artificial objects or some of light-source] is difficult and it can contain simultaneously linear, elliptic, circular or partial polarized components [94]. The whole polarimetric analysis of such objects needs to investigate. At the same time, application of the standard polarization optical elements, for the investigation of the polarization state of each section, it is necessary to take each complete set of different gradient elements, and for this reason it is difficult and labor-intensive to carry out such investigations.

There exist, the optoelectronic and number methods for the polarimetric analysis of objects, but they are difficult and expensive [95, 96]. It is possible to simplify the known methods of polarimetric analysis in above case and instead of the separate anisotropic analysis, phase films will be made which concrete on GB-element based on PVA or PET gradient anisotropic films [93]. In particular, polarization compensators, strictly graduated for the several wavelengths can be constructed. The oriented films created using PVA or PET, corresponding to a given wavelength, will be placed between special optical glasses or quartz plates of definite thickness parallel to each other, and the scale corresponding to the separate wavelength will be located on the conformable section of the GB-element. This element is universal because it replaces simultaneously the anisotropic phase films of all type and it is not necessary to apply different complete sets for the investigation of the different section of the object for polarimetric analysis. If such compensators were applied to microscopes, its geometry sizes will be sufficiently small for the ocular with the focus distance F=25mm to be no more $(10.0\text{-}17.0)\times(5.0\text{-}6.0)\text{mm}^2$ when the total thickness d=1.00mm. The preference of this method, in comparison with the methods of creation of the existent same materials, is: the absence of the moving elements, presence of the graduated scale across the gradient of birefringence;

The real possible application is in the area of polarization compensators for the photo-elastic analysis of the tension effects in hard bodies. Most of all it is possible to prepare such difficult anisotropic films based on GB-elements, with any size, which structurally will be the negative of the tension anisotropic object, i.e., polarization conjugate filter. Similar compensators will give us not only the possibilities of instantly establishing the isochromous and isoclinal effects in the anisotropy of the object, but will be related to the identification of stress in the objects. This is very important in the identification of different faces and objects.

The creation of gradient anisotropic films, such as GB-elements with an optional configuration and a given distribution of polarization gives the possibility of investigation, in the defined conditions and space, the possibility of the biological effects (well-known Haidinger phenomenon) under the influence of the polarized light, on the living organisms (for example, ants, bees, termites, drosophilae, etc.).

At this time, it seems possible to use these in compensators in photonics and photochemistry. It is known that, in photochemistry, studies are being carried out to establish the photosensitivity of a given light-sensitive materials relative to a particular polarizations and the ratio of the values of liner and non-liner Weigart effects [97] in these materials. Such, a strictly graduated GB-element (anisotropic film) provide radiation for the light-sensitive materials, simultaneously, by all type polarization whilst the polarization varies continuously from linear to orthogonal linear, passing through all types of ellipsoid and circular polarizations. On the other head, intensive scientific investigation are carried out in the polarization holography. They are based on the application of the above mentioned Weigert effect [98, 99]. That's why, to-day, it is necessary to predict the polarization characteristics of light waves in holography as well as in photochemistry. Usually, the graduation of the light intensity is fixed; therefore the dynamic range of the process is sufficiently limited, which was connected with the frequency-contrast characteristic of the light-sensitive materials. In the case of fixation of polarization, the conclusive is the fixation of practically infinite variation of polarization, but not the fixation of graduations of intensity. This provides the possibilities of investigation of the holographic and photochemical processes. Hence, it is necessary to ensure in the point of polarization of the difficult special modulation of the light. From this point of view, it is very important to define the strictly graduated GB-elements as standard space polarization modulators. Such elements give us the possibility to achieve the formation of the light-waves with polarization characteristics. For this purpose, in polarization holography these are practically indispensables in obtaining the above mentioned GB-elements with defined parameters relative to the laser lines. The application of anisotropic films such as GB-elements in holography, will considerably expand by the possibilities of the optical information treatment methods.

It is also relevant to the current problem of the creation of miniature spectral devices. Miniature monochromators have already been created, which are based on the application of the interferometric films. These films [100] differs from the others in that the thickness of the multi-layer dielectric system varies steadily across the filter so, that on one side of the filter the maximum of the transparency lies in the red range, and on the other side – in the violet range. But interferential-polarization filters have already been created, with high spectral resolution. These filters allow an exceedingly narrow region of spectral transparency, the width of which area 1/10 of an angstrom or less. The basic node of such a filter is the anisotropic filter, which is placed between the crossed polarizer and the analyzer.

The dependence of the resolution of the interferential-polarization filter transparency on the wave-length could be express by the equation: $T=cos^2 \pi (n_o-n_e)d/\lambda$,where n_o and n_e are the refractive indexes of usual and unusual beam; d-the thickness of the film, λ - wave-length of transmitted light. When $n_o-n_e=\Delta n$ (which determines the anisotropy) varies across the film, according to a definite form, the interferential-polarization filter produces a monochromator, with resolution considerably higher then the resolution of the above. As is known, the interferential - polarization filters have also one shortcoming. Particularly, they differ from the other ones by the existence of the multiple ranks of maximums and minimums (sequential, result from the formula). Therefore such filters will be applied in combination with the usual and interferential filters. Now, we are making the corresponding interferential-polarization dispersion combinatorial filter with the high resolution. The obtained combinatorial system can be placed between the crossed polarizer and analyzer. On spite of this, that intensity is twice diminished on the end of the filter; its spectral resolution is appreciably high, then the resolution of the usual ones. It will be noted, that for the these combinatorial filters, it is necessary to coordinate (reconciliation) the distribution of the spectral transparency of the interferential-polarization dispersion filter with the control of this distribution of the anisotropy in the GB-element.

It is possible to apply the gradient anisotropic films in the luminescence analysis. In particular, the irradiation of luminescence-able dyes dispersed in the oriented polymer films as in matrix, is partly polarized. Moreover, the degree of the luminescence (fluorescence) polarization, substantially depends on the structure of the luminescence-able organic compound, also the degree of orientation of the polymer matrix, on the anisotropy of this materials. In this case, if the anisotropy of the polymer matrix film varies along the axis, i.e. we have the gradient anisotropic matrix; the degree of the luminescence polarization is characterized by a defined distribution, i.e. with the gradient in the same direction. If the degree of the polarization of the luminescence along the x-axis is $P_L(x)$, and the degree of the polarization of the polymer matrix is $P_M(x)[(or\ \Delta n(x)]$, then the relative value $K=P_L(x)/P_M(x)$ or $K=P_L(x)/\Delta n(x)$ together with the intensity of the luminescence and spectral characteristics must be determined by certain properties of the luminescent.

It is possible that the gradient anisotropic films such as GB-elements would be applied successfully in interferometry. It is interesting to carry out the investigations of electro- and magnet-optical, piezo-electrical and the other properties [93] of the above mentioned elements. Recently results have been obtained on this application of GB-films based on the above mentioned polymer materials in polaroscope for investigation of the deformation of transparent polymer materials: definition of the tension at bound of bodies, holes and cuts at the junction of the parts; investigation of the anisotropy and large deformations, heat deformations, residual tensions.

The preference of this method in comparison with the existent methods of creation of the same materials are the presence of the scale of the value of birefringence Δn along the gradient Δn, and the possibility to fix and bind of values of Δn relative to the difficult interferential picture field.

ACKNOWLEDGMENTS

Authors thanks to Prof. J. Aneli for his help.

REFERENCES

[1] Micaelian, A.L. / Optical wave-guides with change-able refractive index // *Optics and spectroscopy,* 1978, vol. 44, pp. 347-370.

[2] Pearson A.D., French W.G., Rawson E.G. // *Appl. Phys. Lett.* 1969, vol. 15, pp. 76-77.

[3] Ohtsuka Y. // *Apll. Phys. Lett.* 1973, vol. 23, p. 247.

[4] Moore D.T. Gradient-index Optic. Review // *Appl. Opt.,* 1980, vol. 19, No 7, pp. 1035-1038.

[5] Buxbinder T.L., Kosiakov B.I., Tukhvatulin A.Sh. / Problems of polymer gradient optics // The Vth All-Union-congress (Former USSR) on polymer optical materials. The abstracts of plenary and stand reports; Jul., 11-14, 1991, pp. 9-10.

[6] Ohtsuka Y., Terao Y./ Studies on the light-focusing plastic rod. IX. Chemical composition of the copolymer rod-diethylen glycol bis allyl carbonate with 2,2,3,3-tetrafluoropropylmethacrylate. // *J.Appl. Polym.Sci.* 1981. 26, pp. 2907-2915.

[7] Oguara Y. / Application of gradient index fiber array to copying machine. // 1981. Kogaku, 10, pp. 111-117.

[8] Koike Y., Hidaki H, Ohtsuka Y. / Plastic axial gradient-index lens // *Appl. Opt.* 1985, vol. 24, No 24, pp. 4321-4325.

[9] Smirnov B.R., Berezin M.P., Rashchupkina I.E. / The application perspectives of polymer materials in lens optics. Polymer optical materials. // *Collection works.* Chernogolovka (Russia) 1989, pp.3-38.

[10] L.P. Asatiani, N.G. Lekishvili, G.M. Rubinstein and W.S. Chagulov, *Functional Classification of Optically Prepared Polymeric Materials*, Tbilisi University Press, Georgia, 1990 [in Russian].

[11] N.G. Lekishvili, G.M. Rubinstain and L.P. Asatiani, *Polymer Waveguides: Obtained, Application, Possibilities.* Tbilisi University Press. 1994 [In Russian].

[12] Nadareishvili L.I., Lekishvili N.G. / Polymer media with the gradient of optical properties. *The controled gradient-formation.* Publishers: "Lega". Tbilisi, 1999 [In Russian].

[13] Tagaya A., Koike Y., / Higly scattering optical transmission polymers for bright display. // *Macromolecular Symposia* Volume: 154, Issue: 1.Date: April, 2000, pp. 73-82.

[14] Koike Y., Ishigure T., Satoh M., Nihei E. / High-speed photonics polymers and its application. // *Pure Appl. Opt.* 7 (March 1998), pp. 201-210.

[15] Marchand E.W. / Gradient index lenses. // *Progr. Opt.,* 1977, vol. 11, pp. 305-307.

[16] Linsky R. / Gradient-index meniscus lens free of spherical aberration. // *J. Opt.* 2000, 2, pp.449-451.

[17] Lekishvili N., Nadareishvili L.,Zaikov G., Khananashvili L. Polymers and polymeric materials for fiber and gradient optics. *VSP* (Utrecht-Boston-Köln-Tokyo), 2002.

[18] Ilin V.G., Karapetiamn G.O., Remizov P.V., Petrovskii G.T., Polianskii M.P. / Optics of gradans // *Sacsseses of photography* 1985, vol. 23, pp 110-121 [In Russian].

[19] N. Lekishvili, L. Nadareishvili, G. Rubinshtein, K. Japaridze / Polymer media with the gradient of optical properties // International conference "Fundamental problems of sciences about polymers" (Dedicated to academician V.A. Kargin's 90th b. d.) In: *book of Abstracts*. 1997, p. C-57.

[20] Ohtsuka Y., Nakomoto Y. / Light focusing plastic rod prepared by photopolymerization of methacrylic ester with vinyl benzoate. //*J. Phys. Lett.,* 1976, vol. 29, pp. 559-561.

[21] Ohtsuka Y., Koike Y., Yamazuki H. / Studies on the light-focusing plastic rod 6: the photopolymer rod of methylmethacrylate with vinyl benzoate. // *Appl. Opt.,* 1981, vol. 20, pp. 280-285.

[22] Koike Y., Kimoto Y., Ohtsuka Y. / Studies on the light-focusing plastic rod 13: photo-copolymerization methylmethacrylate with vinyl ester of aromatic carbozylic acid. // *Appl. Polym. Sci.,*1982, vol. 27, pp. 3253-3264.

[23] Koike Y., Ohtsuka Y. / Studies on the light-focusing plastic rod 15: GRIN-rod by the photocopolymerization of a ternary monomer system. // *Appl. Opt.,* 1983, vol. 22, No 3, pp. 418.

[24] Koike Y., Ohtsuka Y. / Studies on the light-focusing plastic rod 18: Control of refractive-index distribution of plastic radial gradient-index rod by photocopolymerization. // *Appl. Opt.* 1985, vol. 24, No 24, pp. 4316-4320.

[25] Buxbinder T.L., Kosiakov B.I. / About the role of gel-effect at creation of polymer gradient mutually penetrating net structure based on diallilisophtalat - methylmetacrylate // *Visokomolek. Soed.* (Macromolecular compounds, Russia) 1985, vol. 33B (Short communications), No8, pp. 625-627 [In Russian].

[26] Kandelaki S.A., Lezhava B.S., Lekishvili N.G. / The creation method of upper for light-focusing polymer materials // *Author Certificate.* 156196 (USSR) [In Russian].

[27] Kandelaki S.A. PhD thesis "Diffusion formation of the gradient of refractive index in polymer media based on diallyl monomers. Leningrad Cellulose-paper Technological Institute, Leningrad, 1989 [In Russian].

[28] Airapetov I.S., Gachechiladze N.G., Grigoriev S.I., Dianov E.M., Kandelaki S.A. /About polarization properties of polimer gradient pivot // *Kvantovaia elektronika* 1982, No2, pp. 389-390 [In Russian].

[29] Ohtsuka Y., Sugano T. / Studies on the light-focusing plastic rod 14: GRIN-rod of CR-39 (DEGBAK) trifluoroethyl metacrylate co-polymer by a vapor-phase transfer process.// *Appl. Opt.,* 1983, vol. 22, No 3, pp. 413-417.

[30] Tikhonova L.I., Galimov N.B., Kosiakoc V.I. / The method of obtaining of polymer gradans // *Author Certificate.* 1007384 (USSR).

[31] Ginsburg L.I., Krivchenko L.I., Budtov V.P. / The creation method of upper for optical light-focusing elements and fibers // *Author Certificate.* No 722128 (USSR) [In Russian].

[32] Lekishvili N.G., Kandelaki S.A., Asatiani L.P., Tsomaya N.R. / Light-focusing polymer elements based on diallyl esters of meta-carboran dicarbon-acid // *Bull.of the Academy of Sciences of Georgia,* 1995, vol. 151, No1,pp. 249-253 [In Georgian].

[33] Kandelaki S.A., Gogoberishvili K.M., Lekishvili N.G., Chagulov V.S., Khananashvili L.M. / Investigation of diffusion processes fluorine and silicon-containing acrylates into

gel-polymer matrix based on diallyl monomers // *Bull.of the Academy of Sciences of Georgia,* 1989, vol. 134, No3, pp. 553-556 [In Georgian].

[34] Lekishvili N.G., Kandelaki S.A., Gogoberishvili K.M., Chagulov V.S., Khananashvili L.M // Light-focusing polymer elements based on diallyl, fluorine and silicon-containing monomers. // *Book of abstracts, XIV Mendeleev Congress on General and Applied Chemistry.* Tashkent, 1989, p. 24 [In Russian].

[35] Ohtsuka Y., et.al. // *Appl. Phys. Lett.* 1974, vol. 25, No 11, pp. 659-661.

[36] Ohtsuka Y., Terao Y. / Studies on the light-focusing plastic rod IX: Chemical composition of the copolymer rod. Diethylenglycol bis(allylcarbonate) with 2,2,3,3–tetrafluoropropyl- methacrylate // *Appl. Polym. Sci.* 1981, vol. 26, pp. 2907-2915.

[37] Koike Y.,Hidaki H., Ohtsuka Y. / Plastic axial gradient-index lens // *Appl. Opt.* 1985, vol. 24, No 24, pp. 4321-4325.

[38] Budovskaya L.D., Karapetian G.O., Tikhonova L.I., Kosiakoc V.I. / Zhurnal prikladnoi khimii (*Journal of applied chemistry*), 1984, No 5, p. 1730 [In Russian].

[39] Nadareisvili L.I., Lekishvili N.G., Kandelaki S.A., Asatiani L.P., Japaridze K.G., Khananashvili L.M. /Light-focusing optical elements with refractive coefficient gradient based on some organic and organo-element monomers // *Russian Polymer News.* 1999, vol. 4, No3, pp. 4-12

[40] Lekishvili N.G., Kandelaki S.A., Gogoberishvili K.M., Chagulov V.S., Khananashvili L.M./ Polymer selfocs based on systemOCM-2-Styrene-Fluoroalkylmetacrylate // *Bull. of the Academy of Sciences of Georgia.,* 1990, vol. 137, No3, pp. 537-540 [In Georgian].

[41] Lekishvili N.G., Rubinstein G.M., Nadareisvili L.I., GuliashviliT.T., Asatiani L.P/ The creation method of obtaining of light-focusing polymer element // *Patent of Georgia,* No 557. Bull. (10), 1997[In Georgian].

[42] Nadareishvili L.I., Gvatua Sh.Sh., Topuridze N.S. / Investigation of thin-layer selfoc from chlorinate iso-tactic polypropylene // Opticheskii Zhurnal (*Optical Journal*) 1997, vol. 64, No12, pp 69-73 [In Russian].

[43] Rubinstein G.M., Chagulov V.S. / Dypole-phorez effect in polymer // Polymer optical materials. *Collection works.* Academy of Sciences of USSR, Institute of Chemical physics, Chernogolovka. 1989, pp.112-126 [In Russian].

[44] Lekishvili N., Nadareishvili L. / Perspective Polymer Gradient Materials for New Generation Optical Instrument-Making Industry. // *V. A. Kargin's Reading.* Moscow, 2004, January, p. 82.

[45] Galimov N.B., Kosiakov B.I., Tukhvatulin A.Sh., et.al. / Investigation of formation of the refractive index profile at manufacturing of polymer light-focusing elements / *Optica i spectroskopia (Optics and spectroscopy),* 1981, vol. 50, No 3, pp. 546-549 [In Russian]..

[46] Nikiforenko V.S., Kondratovich A.A., Alekseev N.N., et. al / Refractometric analysis of process of homopolymerization of diethyleneglycol-bisallylcarbonate // *Visokomolek. Soed. (Macromolecular comp.)* 1986, vol. 28B (Short communications), No1, pp. 57-62 [In Russian].

[47] Ushakova I.L., Denisov V.M., Lebedeva M.F., et.al. / Study of polymerization of diallylisophtalat by methods od DSC and PMR // *Visokomolek. Soed. (Macromolecular comp.)* 1989, vol.(A)31, No1, pp. 140-146 [In Russian].

[48] Smetanina I.K., Emelianov D.N. / Physical-chemical basis of synthesis and proceeding of polymers. Gorky State University Publishers 1980, pp 56-58 [In Russian].
[49] Report of Scientific-research work, Number of State registration 01840067242, USSR, 1984-1986 (Ginzburg L.I., Krivchenko E.I., Budtov V.P). Leningrad: 1988 [In Russian].
[50] Nikiforenko V.S., Alekseev N.N., Zaitsev I.C. / Radical polymerization of diethylenglycol-bisallyl-carbonat // *Visokomolek. Soed. (Macromolecular comp.),* 1986, vol.(A)28, No10, pp. 2063-2068 [In Russian].
[51] Tukhvatulin A.Sh., Barkovskii B.P., Tikhonova L.I., et.al. / Optimal conditions of diffusion for obtaining of light-focusing elements // *Zhurnal Tekhnicheskoi Khimii (Journal of Technical Chemistry),* 1980, vol.50, No3, pp.1345-1347 [In Russian].
[52] Karapetian G.O., Kosiakov B.I. / Polymeric selfocs: properties and application of optically transparent polymers (Ed.: Shepurev E.I.). // *Book of abstracts All-union* (USSR) School for young scientists and specialists on optical remake of information. 1984, Kiev, pp. 296-298 [In Russian].
[53] Kosiakov B.I., Karapetian G.O. // *Materials of Scientific-technical conference: Properties and application of optically transparent polymers* // Leningrad: 1984 [In Russian].
[54] Lekishvili N.G., Nadareisvili L.I, Kandelaki S.A., Gvatua sh.sh., Topuridze N.S. Japaridze K.G.. / Polymer optical elements with gradient of refractive index based on some organic and organo-element monomers. // Proceedings of the academy of sciences of Georgia. *Chemical Series* . 1999, vol. 25, No1-2, pp. 439-445 [In Georgian].
[55] Budovskai L.D., Ivanova V.I., Kosiakov V.I. / Diffusion of fluoroalkylmetacrylates into for-polymer setting matrix. // *Visokomolek. Soed. (Macromolecular comp.),* 1988, vol. (A)30, No7, pp. 1526-1533 [In Russian].
[56] Kandelaki S.A., Lekishvili N.G., Gogoberishvili K.M., Chagulov V.S., / Investigation of diffusion processes of fluorine- and silicon-containing acrylates into gel-polymer matrix based on diallyl monomers. // *Bull. of the Academy of Sciences of Georgia.,* 1989, vol. 134, No 3, pp. 553-556 [In Russian].
[57] Bukhbinder T.L., Kosiakov B.I., Tukhvatulin A.Sh. / Diffusion molecular change of monomers at formation of gradient structures // *Visokomolek. Soed. (Macromolecular comp.),* 1991, т.32Б (Short communications), No 4, pp. 246-251 [In Russian].
[58] Kandelaki S.A., Lekishvili N.G., Mikadze A.Z., Chagulov V.S., / Diffusion formation of the gradient of refractive index in polymer light-focusing elements based on diallylisophtalat and diethylenglycolbisallylcarbonate // Proceedings of the academy of sciences of Georgia. *Chemical Series* . 1992, vol. 18, No 2, pp. 124-131 [In Russian].
[59] Kosiakov B.I., Tukhvatulin A.Sh.,.et. al. / Zhurnal Tekhnicheskoi Fiziki (*Journal of technical physics*), 1995, vol. 65, No 4, pp.46-56 [In Russian].
[60] Bukhbinder T.L., Dremina E.A., Kosiakov B.I., et.al. / Essential regulations of formation of refractive index gradient in planar lenses. // *Zhurnal Tekhnicheskoi Fiziki (Journal of technical physics),* 1997, vol. 67, pp. 83-87 [In Russian].
[61] Nadareishvili L.I, Bakuradze R.Sh., Pavlenishvili I. I. / About question of morphology and staratification of iso-refraction in GRIN-elements. // *Georgian Engineering News,* 2002, No 2, pp. 75-78 [In Russian].
[62] Fadeev E.A., Kosiakov B.I., Tukhvatulin A.Sh.,.// *Zhurnal Tekhnicheskoi Fiziki (Journal of technical physics),* 1998, v. 68, No10, pp. 70-73 [In Russian].

[63] Lekishvili N., Rubinstein G., Asatiani L.,Tsomaia N., Andguladze N., Kandelaki S. / Refractometric investigation of polymer gradient optical media based on some organic monomers.// Proceeding of the Academy of Sciences of Georgia. *Chemical Series*. 1993, vol.19, No 3-4. pp.

[64] Rubinstein G., Lekishvili N., Kandelaki S., Khananashvili L. / Refractometric investigation of polymer gradient optical media based on some organic and element-organic monomers // *Intern. J. Polym. Mater.* 1998, vol.41, pp. 79-88.

[65] Chen W.C., ChenI.H., Yang S.Y. et. al. // *Photonic and optoelectronic polymers*. By Ed. Samson A. Lenkhe, Keneth E.Wynnei. Chapter 6. A.C.S. Washington, DC, 1997.

[66] Jui-Hsiang Liu and Hung-Tsai Liu. / Gradient refractive-index optical rod prepared from methacrylate derivatives. // *Optic Letters,* 1997, vol.22, No 10, pp. 668-670.

[67] Sheng Pin Wu, Eisuke N., Koike Y. // Large radial graded-index polymer. / *Appl. Opt.* 1996, vol. 35, No 1, pp. 28-32.

[68] Koike Y., Ishigure T., Eisuke N. / High-Bandwidth graded-Index polymer Optical Fiber. //*Journal of lightwave technology,* 1995, vol.13, No 7, pp. 1475-1489.

[69] Nadareishvili L.I., Topuridze N.S, Gvatua Sh.Sh., Lekishvili N.G., Japaridze K.G. / Creation of the polymer media with the given gradient of refractive index // Proceedings of the academy of sciences of Georgia. *Chemical Series*. 1999, vol. 25, No 3-4, pp 273 [In Georgian].

[70] Nadareishvili L., Lekishvili N., Japaridze K., Khananashvili L. / Polymer media with gradient of refraction parameter // *Russian Polymer News*. 1997, vol.2, No3. pp. 30-35.

[71] Karapetian G.O., Kosiakov B.I., Moriganov A.N. / Gradient planar structures // *Book of abstracts of 5th All-union (USSR) on polymer optical materials.* Leningrad. June, 11-14. 1991 [In Russian].

[72] Hablmen D.P. / Gradient refractive index optical lenses. Patent USA, No 348608. G02B/00. *Filed Mar.* 14. 1966. Patented Dec., 30, 1969.

[73] Nadareishvili L.I. / Obtaining of multi-canal light-focusing matrix by diffusion-copolymerization of monomers // Proceedings of the academy of sciences of Georgia. *Chemical Series*. 2000, vol. 26, No 3-4, pp. 157-161 [In Georgian].

[74] Nadareishvili L.I. / Regulation of gradient-formation process on the spherical surface // *Journal "Intellect"*. 2000, No 1, pp. 62-67 [In Georgian].

[75] Nadareishvili L.I., Gvatua Sh.Sh., Lekishvili N.G., Japaridze K.G. / Method of obtaining of selfocs by-gamma-irradiation of iso-tactic polypropylene // Georgian Engineering News.1999, No, 4, pp. 83-86 [In Georgian].

[76] Nadareishvili L., Gvatua Sh., Blagidze I., Topuridze N., Japaridze K., Lekishvili N., / Change of some optical properties in polymeric transparent films and formation of the GRIN (Gradient Refractive Index)-media under action of gamma-radiation. // *Russian Polymer News*. 2002, vol. 7, No3. pp. 32-37.

[77] Nadareishvili L.I., Skhirtladze I.A., Japaridze K.G. /The method for obtaining of gradient elements. // *Author certificate*, No 1808733 (Rusian Federation) B29D11/00. Bull. No 2, 1993 [In Russian].

[78] Nadareishvili L.I., Skhirtladze I.A., Japaridze K.G. / The method for creation of light-conducting elements. // *Patent of Georgia*, No 203, B29D11/00, Bull. 5, 1995 [In Georgian].

[79] Nadareishvili L.., Topuridze N., Akhobadze V., Japaridze K. / The method for making of lens // *Patent of Georgia,* No 887, B29D11/00, Bull. No 7, 1997 [In Georgian].

[80] Nadareishvili L.., Topuridze N., Akhobadze V., Japaridze K. / The method for obtaining of planar polymer lens // *Patent of Georgia,* No 996, B29D11/00, Bull. No7, 1997 [In Georgian].

[81] Nadareishvili L.., Topuridze N., Akhobadze V.,Japaridze K. / The method for making of lens // *Patent of Georgia,* No 997, B29D11/00, Bull. No7, 1997 [In Georgian].

[82] Nadareishvili L.., Topuridze N., Blagidze I., Japaridze K. / Device for obtaining given gradient of refractive index in polymers / *Patent of Georgia,* No P1963, B29D11/00, Bull.No13, 1999 [In Georgian].

[83] Nadareishvili Levan and Lekishvili Nodar / Fabrication of Gradient Refractive Index Materials with Controlled Gradient Formation // POLYCHAR WORLD FORUM ON POLYMER APPLICATIONS AND THEORY, POLYCHAR-9. University of North Texas, Denton, Program and Book of Abstracts, 2001, January 9-12 (P.62).

[84] Nadareishvili L., Gvatua Sh., Lekishvili N., Japaridze K. / Polymer GRIN-media formation: non-traditional approaches. // In *"Chemistry and Biochemistry on the Leading Edge.* Nova Sci. Publ. New York. Ed. G. Zaikov. 2002, pp.1-19.

[85] Nadareishvili L., Topuridze N., Favlenishvili I., Japaridze K. / Device for obtaining given gradient of refractive index in polymers / *Patents of Georgia,* No P 2574, P 2575, P 2576; B29D11/00, Bull. No 21, 2001[In Georgian].

[86] Nadareishvili Levan and Lekishvili Nodar / Fabrication of Gradient Refractive Index Materials with Regulated Gradient Formation // POLYCHAR WORLD FORUM ON POLYMER APPLICATIONS AND THEORY, POLYCHAR-9. University of North Texas, Denton, Program and Book of Abstracts, 2001, January 9-12 (9-p.62).

[87] Nadareishvili L., Nakaidze J., Japaridze K / The method for fabrication planar lenses from powder-like components // Proceeding of the Academy of Sciences of Georgia. *Chemical Series,* 1995, v.21, No4, p. 771 [In Georgian].

[88] Nadareishvili L., Gvatua Sh., Lekishvili N., Japaridze K /. POLYCHAR WORLD FORUM ON POLYMER APPLICATIONS AND THEORY, POLYCHAR-8. University of North Texas, Denton (USA). Program and Book of Abstracts, 2000, January, 11-14, p.60;

[89] Nadareishvili L.I. and Gvatua Sh.Sh. International optical congress, Saint-Petersburg (Russia), 2000, October, 16-20.

[90] Nadareishvili L., Gvatua Sh., Zaikov G., Japaridze K. / GB-optics-a new direction of gradient optics // *Journal of the Applied Chemistry,* 2004 (In press).

[91] Nadareishvili L.I., Gvatua Sh.Sh., Topuridze N.S., Blagidze I.M., Sharashidze L.K., Pavlenishvili I.I., Gvatua N.Sh., Japaridze K.G. / Investigation of polymer optical elements with the gradient of birefringence // *Georgia Chemical Journal.* 2003, vol. 3, No 3, pp. 227-230 (In Russian).

[92] Nadareishvili L., Gvatua Sh., Topuridze N., Blagidze I., Sharashidze L., Pavlenishvili I./ Obtaining and investigation of polymer films with birefringence gradient. // *Proceedings of the Institute of Cybernetics of Georgian Academy of Sciences*. 2002, vol. 2, No 1-2, pp. 189-197.

[93] Nadareishvili L / The method of obtaining and investigation polymer films with the given gradient of birefringence. // *Georgian Engineering News*, 2001, No 2, pp. 73-76.

[94] Wadosanidze Z., Nadareishvili L., Lekishvilli N., Gvatua Sh. // *Georgian Chemical Journal.* 2004, vol. 4, No 3, 270-272.

[95] Hide W., Tubbs E.F., Koester C.J., // *J. Journ. Opt. Soc. Amer.* 1959, 49, p.513.

[96] Japhiashvili V.P., Korol A.N. // *Polarimetric Atlas of the Moon,* Tbilisi, Mecniereba, 150, 1982.
[97] Weigert F. / Uber Einen Neuen Effekt Der Strahlung in Lichtempfindlichen Shichten // *Verhardl. Deutschen Physik.* 1919, Bd. 21, 479-483.
[98] Kakichashvili Sh. D. / *Polarization Holography,* Leningrad, 1989, p. 141.
[99] Wardosanidze Z. // *Proceedings of the Institute of Cybernetics* , 2002, No1-2, pp. 205-209.
[100] Zaidel A.N., Ostrovskaia G.V., Ostrovski Y.N. // *The Technique and Practice of the Spectroscopy,* M., Nauka, 1976, p.392.

In: Order and Disorder in Polymer Reactivity ISBN 1-60021-145-3
Editors: G. E. Zaikov and B. A. Howell, pp. 115-124

Chapter 8

EFFECT OF SOLVENT (PLASTICIZERS) ON PVC DEGRADATION

G. E. Zaikov
N.M. Emanuel Institute of Biochemical Physics,
Russian Academy of Sciences
4 Kosygin Str., Moscow 119991, Russia
chembio@sky.chph.ras.ru

ABSTRACT

In Both plasticized (semi-rigid and flexible) PVC materials as well as PVC in solutions, the rate of their thermal degradation and effective stabilization are caused by essentially different fundamental phenomena in comparison to aging of PVC in absence of the solvent.

Both structure and macromolecular dynamics render the significant influence on its stability, i.e. chemical nature of the solvent (plasticizer), its basicity, specific and non-specific solvation, degree of PVC in a solution (solubility), segmental mobility of macromolecules, thermodynamic properties of the solvent (plasticizer), formation of associates, aggregates, etc.

The chemical stabilization of PVC plays a less significant role. The effect of above factors on stability (behaviour) of semi-rigid and flexible PVC will be done on quantitative level.

It will be described effect of "echo"-type of stabilization on the stability of PVC in the presence of plasticizers. If we would like to have stable material from PVC we should make stabilization of plasticizers as more reactive chemical compounds.

Key words: plasticizers, solvent, basicity, specific and non-specific solvation, solution, solubility, segmental mobility of macromolecules, thermodynamic properties, associates, aggregates.

1. Effect of Solvent

At PVC's degradation in solution, one of the basic reasons of change of the process kinetic parameters is the nucleophilic activation of a PVC's dehydrochlorination reaction. The process is described by E_2 mechanism [1-3]. Thus, there is a linear dependence between PVC's thermal dehydrochlorination rate and parameter of solvent's relative basicity B cm^{-1} (Fig. 1) [1-3]. (The value B cm^{-1} is evaluated by shift of a characteristic band OH of phenol at $\lambda = 3600$ cm^{-1} in an IR-spectrum at interaction with the solvent [4]). It is essentially important that the rate of PVC's dehydrochlorination in the solvents with relative basicity $B > 50$ cm^{-1} was always above, than the rate of PVC's dehydrochlorination without the solvent, while when $B < 50$ cm^{-1}, PVC's desintegration rate was always less, than at it's destruction without the solvent. The revealed dependence $V_{HCl} = f(B)$ is described by the equation:

$$V^*_{HCl} = V_{HCl} + k(B - 50) \quad (1)$$

An inhibition of PVC's disintegration in the solvents with basicity $B < 50$ cm^{-1} is very interesting and practically important phenomenon. It has received the name "solvatational" stabilization of PVC. Let's notice, however, that ignoring of the fact that PVC solutions even at low concentration (2 wt. %) do not represent solutions with isolated macromolecules but rather with structured systems, results that in a number of cases a deviation from linear dependence of PVC dehydrochlorination rate of the solvent basicity B cm^{-1} is observed. In particular, an abnormal behavior of PVC is observed at destruction in certain ester-type solvents (plasticizers) (Fig. 1, points 25 - 28), that apparently caused by structural changes of macromolecules. This was never before taken into account at work with PVC in solutions.

It was revealed quite unexpectedly that not only interaction "polymer – solvent", but also interaction "polymer – polymer" in solutions provide significant influence on rate of PVC disintegration. It's known that structure and properties of the appropriate structural levels depend from conformational and configurational nature of macromolecules, including a supermolecular structure of the polymer, which in turn determines all basic (both physical and chemical) characteristic of polymer.

"Polymer – polymer" interaction results to formation of structures on a supermolecular level. In particular, as getting more concentrated the PVC-solvent system consistently passes a number of stages from isolated PVC macromolecules in a solution (infinitely diluted solution) to associates and aggregates from macromolecules in a solution. At the further increase of PVC concentration in a solution formation of spatial fluctuational net with structure similar to a structure of polymer in the block occurs.

When polymer's concentration in a solution increases, the rate of PVC's dehydrochlorination reaction changes as well, and various character of influence of the solvent on a PVC disintegration rate in solution is observed depending on a numerical value of basicity parameter B cm^{-1} [5-10]. If the relative basicity of employed solvents was $B > 50$ cm^{-1}, the polymer's degradation rate decreases when its concentration increases. If a basicity of the employed solvents was $B < 50$ cm^{-1}, the polymer's degradation rate increases with increased concentration of a polymer. In all cases the rate of HCl elimination from a polymer has a trend in a limit to reach values of PVC dehydrochlorination rate in absence of the solvent $V_{HCl}^{PVC} = 5*10^{-8}$ (mol HCL/mol PVC)/s. (Fig. 2).

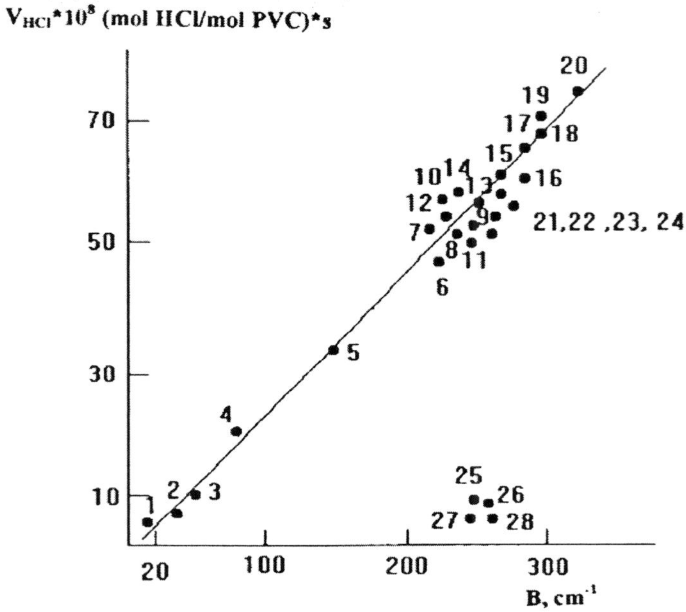

Figure 1. Influence of the solvent's basicity on the rate of thermal dehydrichlorination in solution: 1 – *n*-dichlorobenzene, 2 – *o*-dichlorobenzene, 3 – naphthalene, 4 – nitrobenzene, 5 – acetophenone, 6 – benzonitrile, 7 – di-*n*-(chlorophenyl-chloropropyl) phosphate, 8 – triphenylphosphite, 9 – phenyl-*bis*-(β-chloroethyl) phosphate, 10 – tri-(*n*-chlorophenyl phosphate), 11 – 2-ehtylhexylphenyl phosphate, 12 – tricresyl phosphate, 13 – cyclohexanone, 14 – phenyl-*bis*-(β-chloropropyl) phosphate, 15 – tri-β-chloroethyl phosphate, 16 - tri-β-chloropropyl phosphate, 17 – di-2-(ethylhexyl) phosphate, 18 – 2-ethylhexylnonyl phosphate, 19 – tri-(2-ethylhexyl) phosphate, 20 – tributyl phosphate, 21, 25 – dibutyl phthalate, 22, 26 – di-2-ethylhexyl adipinate, 23, 27 – dioctyl phthalate, 24, 28 – dibutyl sebacinate. Concentration of PVC in solution: 1-24 – 0.2 wt. %, 25 - 28 – 2 wt. %; 423 K, under nitrogen.

Equation (1) turns into an equation (2) if to take into account that the PVC's degradation rate is determined not only by parameter of relative basicity of the solvent B, but also by its concentration in a solution (C, mol PVC/L), as well as by degree of "polymer - polymer" interaction (degree of macromolecules structurizarion in a solution $\Delta C = /C\text{-}C_0/$, where C_0 - concentration of a beginning of PVC macromolecules association in a solution):

$$V_{HCl} = V^0_{HCl} + A_1 /(C + /\Delta C/ + d_1)(B - 50) \tag{2}$$

Here factor $A_1 = (0.8 \pm 0.2)*10^{-9}$ (mol HCL/mol PVC)/s; d_1 - dimensionless factor reflecting interaction "polymer – solvent" ($d_1 = 0.5 \pm 0.25$). The deviation from the moment of a beginning of macromolecules association in a solution is taken on the absolute value, since it can change in both directions to more concentrated and more diluted solutions of a polymer.

Equation (2) well describes a change of PVC's thermal dehydrochlorination rate of its concentration in a solution in view of parameter of relative basicity of the solvent B, irrespective of the chosen solvent (Fig. 3).

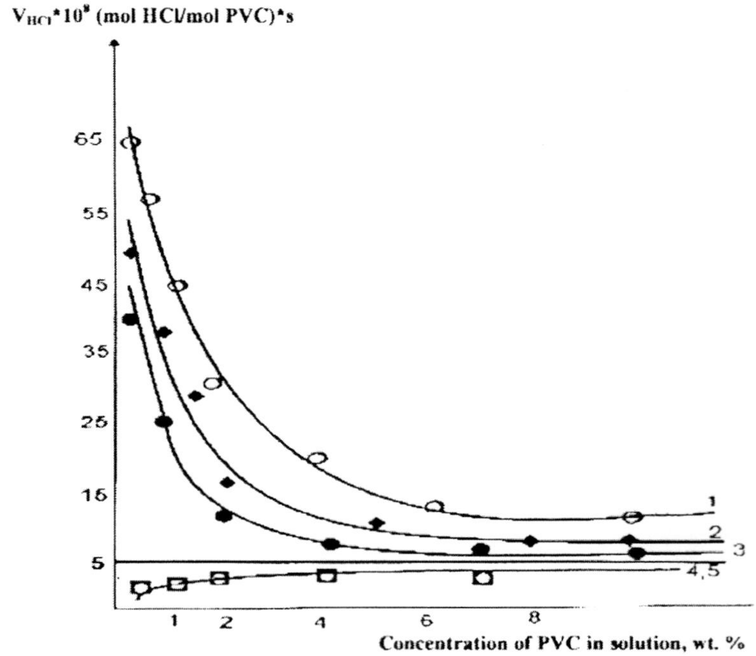

Figure 2. A change of PVC's dehydrochlorination rate of its concentration in a solution: 1 – cyclohexanol, 2 – cyclohexanone, 3 – benzyl alcohol, 4 – 1, 2, 3 – trichloropropane, 5 – *o*-dichlorobenzene, 6 – no solvent; 423 K, under nitrogen.

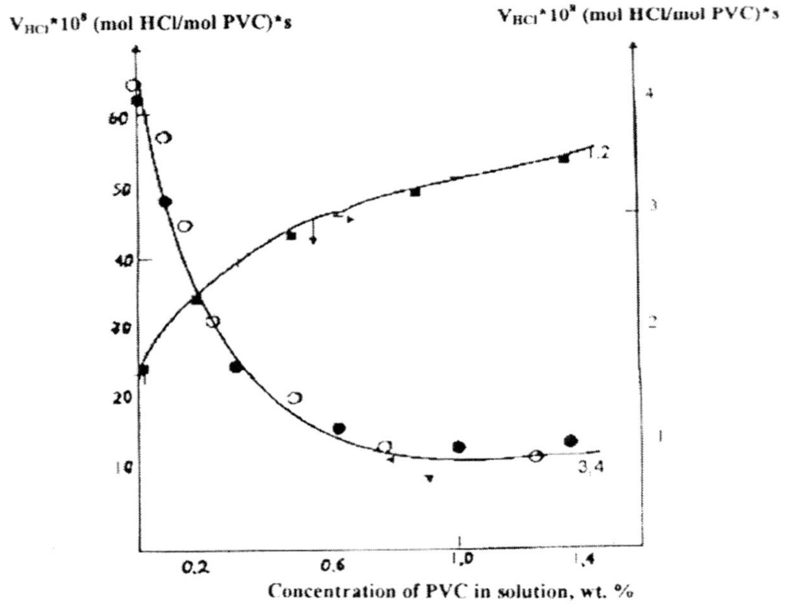

Figure 3. A change of PVC's dehydrochlorination rate from its concentration in a solution: 1, 2 – 1, 2, 3 – trichloropropane, 3, 4 – cyclohehanol, 1, 3 – experimental data, 2, 4, - calculated data with equation (4) at $A_1=10^{-9}$ and $d_1=0.8$ and 0.7 correspondingly; 423 K, under nitrogen.

The observable fundamental effect has the significant importance at production of plasticized (in particular, by esters) materials and products made from PVC. Despite of very high basicity of ester-type plasticizers (B = 150 cm^{-1}) in an interval of PVC concentration in solutions more than 2 %, a noticeable reduction of PVC degradation rate is observed (Fig. 1, curve, points 25 - 28), *i.e.* on – essence, stabilization of PVC occurs. This effect is caused by formation of dense globules, associates, *etc.* in the system PVC - plasticizer. Practically this allows to create economic formulations of plasticized materials from PVC with the very little contents of metal-containing stabilizers - HCl acceptors, or without their use at all.

Temperature is very important on formation of the heterophase system. Even at low concentration of PVC in ester-type plasticizer (for example, in dioctyl phthalate at C> 0.1 mol/L) the true solutions are formed only at temperatures above 400 K. Globular structure of suspension PVC and formation of associates retain at temperatures up to 430 - 445 K. In other words, PVC at plasticization, is capable to keep its structural individuality on a supermolecular level, which is formed during polymer's synthesis. Specifically in these conditions the ester-type plasticizer behaves not as a highly-basic solvent, but as a stabilizer at PVC's thermo-degradation due to formation of associates *etc.* This leads to a reduction of stabilizer's amount, extension of exploitation time of materials and products, *etc.*

It is necessary to note that the change of PVC's degradation rate at association of macromolecules is the general phenomenon and does not depend on how it was achieved. In particular, similar (as well as at concentration of PVC solutions (Fig. 2, 3)), character of change of PVC dehydrochlorination rate in a solution is observed, if the change of PVC's structural - physical condition in a solution is reached upon addition even chemically inert non-solvents, for example, hexane, decane, undecane, polyolefines, polyethylene wax, *etc.* [6, 9-12] (Fig. 4). It is interesting to note that the degree of relative change of PVC disintegration rate under action of the second inert polymer (non-solvent) is much higher, than at concentration of a PVC solution, especially in case of use of the low-basic solvents (trichloropropane, dichlorobenzene – a result of formation of more dense formations on a supermolecular level, corresponding associates and aggregates, thanks to whom there is a significant change of a PVC destruction rate. The more contents of non-solvent (including an inert polymer) in a blend and lower thermodynamic compatibility of components in a solution, the more structural formation takes place in a solution, including one at the presence of a polymer blends (associates, aggregates). Formation of a fluctuational net with participation of macromolecules is probable. Since the reason of change of PVC thermal dehydrochlorination rate in case of its blends with chemically inert thermodynamically incompatible polymers is the same, as at concentration of a PVC solution (structural - chemical changes of a polymer in a solution), the parameters determining the rate of PVC disintegration, will be, obviously, similar. Therefore, at consideration of PVC thermal destruction a concentration of the second polymer in a blend with PVC, as well as a degree of its thermodynamic affinity to PVC have to be taken in account in addition to an influence of polymer's concentration in a solution, basicity of the solvent B cm^{-1} and forces of interaction "polymer – solvent". In view of these factors the equation (2) turns into an equation (3):

$$V_{HCl} = V^{0}_{HCl} + (A_1 / c + /(c + |\Delta c| + d_1 + \alpha n))(B - 50) + A_1 / B)(d_2 \alpha^n / c) \qquad (3)$$

where α - fraction of the second polymer, varying from 0 to 0.99; n – dimensionless parameter describing a degree of thermodynamic affinity of PVC to the second polymer and varying from zero (in a case of a complete thermodynamic compatibility of the components) up to certain value equals ~10 (in a case of a complete thermodynamic incompatibility of the polymers). Dimensionless coefficient d_2 reflects interaction of the second polymer with the solvent. At destruction of PVC in a blend with poly(ethylene) in a solution of dichlorobenzene, trichloropropane, and cyclohexanol it equals 2.5±0.1.

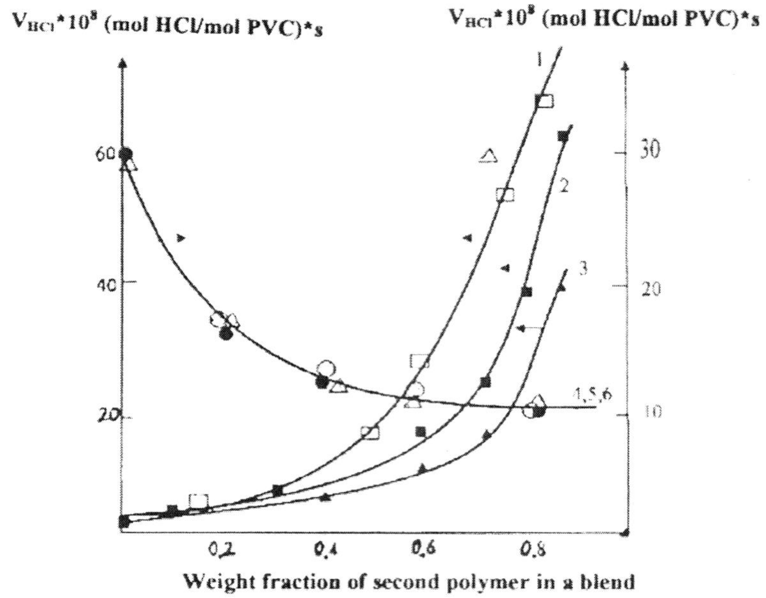

Figure 4. A change of PVC thermodegradation rate of the contents of the secong inert polymer in solution of trichloropropane (1, 3), dichlorobenzene (2), and cyclohexanol (4 – 6) for blends of PVC with poly(ethylene) (1, 4), poly(propylene) (2, 5), and poly(isobutylene) (3, 6); 423 K, under nitrogen.

Observable changes of PVC thermal disintegration rate under action of second thermodynamically incompatible with PVC polymer (or owing to an increase of PVC concentration in a solution) are caused by a displacement of the solvent from macromolecular globules of PVC with transformation to the structure, which it has in absence of the solvent. This evokes unexpected effect of "the solvent action" (a delay or an acceleration depending on the solvent's basicity B cm^{-1}) in relation to PVC's thermal disintegration. The solvent's displacement, which accelerates PVC's disintegration (B > 50 cm^{-1}), results to easing of its interaction with PVC and leads to a delay of process of HCl elimination from macromolecules, *i.e.* to stabilization. This occurs both in a case of concentration of PVC's solutions, and in case of addition of second polymer, which is thermodynamically incompatible with PVC. In the solvents slowing down PVC's disintegration (B < 50 cm^{-1}) by virtue of low nucleophilicity, an effect of the solvent displacement and the easing of its influence on PVC results in an opposite result – an increase of HCl elimination rate from PVC upon of increase of its concentration in a solution or at use of chemically inert non-solvent. It is obvious that irrespective of the fact how to make changes in PVC's structure in a solution – by increase of its concentration in a solution or by addition of second thermodynamically incompatible with PVC chemically inert non-solvent - the varying structural - physical condition of the polymer results in a noticeable change of its thermal dehydrochlorination rate in a solution. These effects are caused by structural - physical changes in system polymer - solvent, and previously unknown phenomena can be classified as structural - physical stabilization (in case of a reduction of gross - rate of PVC disintegration in highly-basic solvents at $B > 50\ cm^{-1}$) and, respectively, structural - physical antistabilization (in case of increase of gross - rate of PVC disintegration in low-basic solvents with $B < 50\ cm^{-1}$).

2. "ECHO" - STABILIZATION OF PVC

At last, it is necessary to specify to one more appreciable achievement in the field of aging and stabilization of PVC in a solution. In real conditions the basic reason of the sharp accelerated aging of plasticized materials and products is the oxidation of the solvent by oxygen of air (Fig. 5, curve 3).

$$RO_2^{\bullet} + RH \xrightarrow{K_2} ROOH + R^{\bullet}$$

$$ROOH \xrightarrow{K_3} RO^{\bullet} + HO^{\bullet}$$

$$RO_2^{\bullet} + RO_2^{\bullet} \xrightarrow{K_6} \text{inactive products}$$

Peroxides, formed at oxidation of ester-type plasticizers, initiate disintegration of macromolecules. In these conditions the rate of PVC destruction can increase in two and more orders of magnitude and is determined by oxidizing stability of the solvent to oxygen - parameter $K_{ef} = K_2 * K_3^{0.5} * K_6^{-0.5}$. Then higher an oxidizing stability of the solvent (in particular, ester-type plasticizer), at which's presence a thermooxidative disintegration of PVC occurs, then lower its degradation rate and longer an exploitation time of semi-rigid and flexible materials on a basis of PVC [13-18]. An inhibition of process of the solvent's oxidation (including plasticizers) due to of incorporation of stabilizers – antioxidants or their synergistic compositions slows a thermo-oxidative disintegration of PVC in a solution (Fig. 5, curve 5).

At effective inhibition of the ester-type plasticizers' oxidation by oxygen of air the rate of PVC thermo-oxidative destruction in their concentrated solutions is getting closer to the rate of polymer's disintegration, what is characteristic for its thermal destruction at plasticizer's (solvent's) presence, *i.e.* slower, than PVC's desintegration without a solvent. This occurs due to a structural - physical stabilization. In these cases an inhibition of reaction of the solvent's oxidation at use of "echo" – type stabilizers - antioxidants causes PVC's stabilization (Fig. 5, curve 5). This fundamental phenomenon of PVC's stabilization in a solution at its thermo-oxidative destruction has received the name of an "echo" - stabilization of PVC [2, 15, 16].

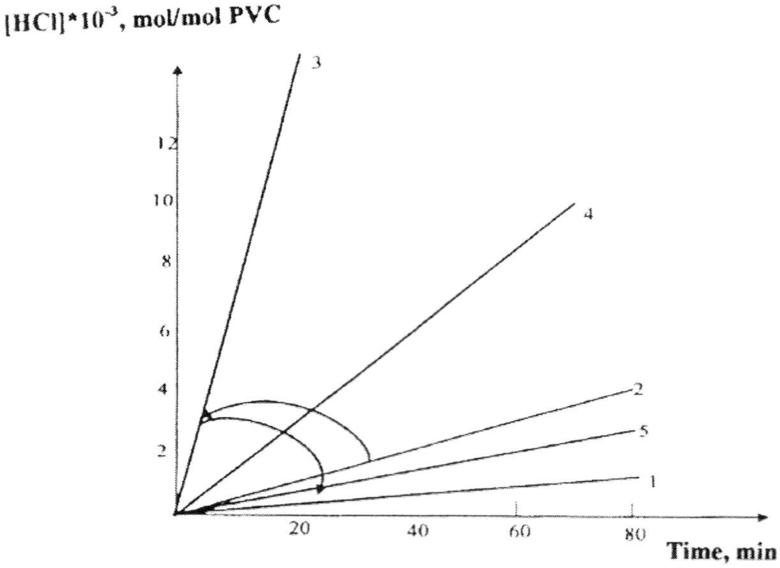

Figure 5. "Echo"-stabilization of PVC. Elimination of HCl during thermo- (argon) (1, 2) and thermo-oxidative (air) (3 – 5) destruction of PVC in solution of dioctyl sebacinate: 1 – 4 unstabilized PVC, 5 – PVC, stabilized with diphenylpropane (0.02 wt. %) – "echo" stabilization; 2, 4 – PVC with no solvent; 448 K.

3. Tasks for a Future

Thus, a creation of high-quality and economic semi-rigid and flexible materials and products on a basis of PVC, including ones where solvents are employed, require the specific approach, essentially differing from principles of manufacture of rigid materials and products from PVC. In particular, account and use of the fundamental phenomena: solvatational, structural - physical and "echo" - stabilization of a polymer in a solution.

As to paramount tasks of fundamental and applied research in the field of PVC's manufacture and processing in the beginning of XXI century, obviously they are following:

- Manufacture of an industrial PVC, not containing of labile groups in a backbone. It will provide drastic increase of an intrinsic stability of polymeric products, possibility of PVC processing with the minimal contents or in total absence of stabilizers and other chemicals - additives and opportunity of creation of materials and products on a PVC basis with the essentially increased service life-time;
- Wide use of the latest achievements in area of destruction and stabilization of PVC, both at presence and absence the solvents. First of all, phenomena of chemical, solvatational, structural - physical, self- and "echo" - stabilization of PVC will allow to create rigid, semi-rigid and flexible (plasticized) materials and products with the minimal contents of chemicals - additives and increased life-time of their service at exploitation in natural and special conditions;

- The use of non-toxic, non-flammable products which do not emit toxic and other poison gaseous and liquid products at elevated temperature at manufacture of materials and products from PVC;
- Complete slimination of all toxic and even low-toxic (particularly compounds based on Pb, Cd, Ba, *etc.*) chemicals – additives from all formulations;
- Search of non-toxic and highly effective inorganic chemicals - additives, first of all, stabilizers of a zeolite - type, modified clays, *etc.*

At the same time new "surprises" may be expected, which undoubtedly will be presented us by this outstanding polymer - puzzle, a plastic's "working horse" for many decades. Certainly it will give new stimulus in development of scientific bases and practical development with opening of new pathways, conducting to essential delay of PVC's ageing in natural and special conditions at reduction of amounts of the appropriate chemicals - additives, down to their complete elimination [18-25].

REFERENCES

[1] Minsker, K. S.; Abdullin, M. I.; Manushin, V. I.; Malyshev, L. N.; Arzhakov, S. A. *Dokl. Akad. Nauk SSSR* 1978, *242(2)*, 366-368.

[2] Minsker, K. S. *Polymer Yearbook*; Pethrick, A., Ed.; Acad. Publ.: Harwood, 1994; p 229-241.

[3] Minsker, K. S.; Kulish, E. I.; Zaikov, G. E. *Vysokomol. Soedin.* 1993, *35B(6)*, 316-317.

[4] Palm, V. A. *Osnovy kolichestvennoi teorii organicheskikh reaktsii*; Khimiya: Leningrad, 1977, p 114.

[5] Kolesov, S. V.; Kulish, E. I.; Minsker, K. S. *Vysokomol. Soedin.* 1994, *36B(8)*, 1383-1384.

[6] Kolesov, S. V.; Kulish, E. I.; Zaikov, G. E.; Minsker, K. S. *Russian Polym. News* 1997, *2(4)*, 6-9.

[7] Kulish, E. I.; Kolesov, S. V.; Minsker, K. S. *Bashkirskii Khimicheskii Zhurnal* 1998, *56(2)*, 35-37.

[8] Kulish, E. I.; Kolesov, S. V.; Minsker, K. S.; Zaikov, G. E. *Vysokomol. Soedin.* 1998, *40A(8)*, 1309-1313.

[9] Kolesov, S. V.; Kulish, E. I.; Zaikov, G. E.; Minsker, K. S. *J. Appl. Polym. Sci.* 1999, *73(1)*, 85-89.

[10] Kulish, E. I.; Kolesov, S. V.; Minsker, K. S.; Zaikov, G. E. *Chem. Phys. Report* 1999, *18(4)*, 705-711.

[11] Kulish, E. I.; Kolesov, S. V.; Akhmetkhanov, R. M.; Minsker, K. S. *Vysokomol. Soedin.* 1993, *35B(4)*, 205-208.

[12] Kulish, E. I.; Kolesov, S. V.; Minsker, K. S.; Zaikov, G. E. *Int. J. Polym. Mater.* 1994, *24(1-4)*, 123-129.

[13] Martemyanov, V. S.; Abdullin, M. I.; Orlova, T. E.; Minsker, K. S. *Neftekhimiya* 1981, *21(1)*, 123-129.

[14] Minsker, K. A.; Abdullin, M. I.; Zueva, N. P.; Martemyanov, V. S.; Teplov, B. F. *Plast. Massy* 1981, *9*, 33-34.

[15] Minsker, K. S.; Abdullin, M. I. *Dokl. Akad. Nauk SSSR* 1982, *263(1)*, 140-143.
[16] Minsker, K. S.; Zaikov, G. E. *Chemistry of chlorine-containing polymers: synthesis, degradation, stabilization*; NOVA Science: Huntington, 1999; p 295.
[17] Minsker, K. S.; Lisitsky, V. V.; Zaikov, G. E. *Vysokomol. Soedin.* 1981, *23A(3)*, 498-512.
[18] Minsker, K. S.; Kolesov, S. V.; Yanborisov, V. M.; Berlin, Al. Al.; Zaikov, G. E. *Vysokomol. Soedin.* 1984, *26A(5)*, 883-899.
[19] Minsker, K. S.; Lisitsky, V. V.; Zaikov, G. E. *J. Vinyl Technol.* 1980, *2(4)*, 77-86.
[20] Minsker, K. S.; Kolesov, S. V.; Zaikov, G. E. *Degradation and Stabilization of Vinyl chloride based polymers*; Pergamon Press, 1988.
[21] Minsker, K. S.; Kolesov, S. V.; Zaikov, G. E. *J. Vinyl Technol.* 1980, *2(3)*, 141-151.
[22] Minsker, K. S.; Kolesov, S. V.; Zaikov, G. E. *Vysokomol. Soedin.* 1987, *23A(3)*, 498-512.
[23] Minsker, K. S.; Kolesov, S. V.; Yanborisov, V. M.; Adler, M. E.; Zaikov, G. E. *Dokl. Akad. Nauk SSSR* 1983, *268(6)*, 1415-1419.
[24] Kolesov, S. V.; Minsker, K. S.; Yanborisov, V. M.; Zaikov, G. E.; Du-Jong, K.; Akhmetkhanov, R. M. *Plast. Massy* 1983, *12*, 39-41.
[25] Kolesov, S. V.; Steklova, A. M.; Zaikov, G. E.; Minsker, K. S. *Vysokomol. Soedin.* 1986, *28A(9)*, 1885-1890.

In: Order and Disorder in Polymer Reactivity
Editors: G. E. Zaikov and B. A. Howell, pp. 125-135
ISBN 1-60021-145-3

Chapter 9

WATERSOLUBLE POLYMERS BASED ON N,N-DIALLYL-N'-ACETYLHYDRAZINE

A. I. Vorob'eva*[*,1], *M. N. Gorbunova*[2], *S. I. Kuznetsov*[1], *R. R. Muslukhov*[1], *S. V. Kolesov*[1], *A. G. Tolstikov*[2] and *Yu. B. Monakov*[1]

[1] Instutute of Organic Chemistry, Ufa Scientific
Centre of Russian Academy of Sciences
October pr., 71, Ufa, Russia, 450054
[2] Institute of Technical Chemistry, Ural Branch
of Russian Academy of Sciences
Lenin st., 13, Perm, Russia, 614990

Activity of *N,N*-diallyl-*N'*-acetylhydrazine in reactions of radical homo- and copolymerization has been studied. It was determined, that *N,N*-diallyl-*N'*-acetylhydrazine does not homopolymerize according to free-radical mechanism, but copolymerize with N-vinylpyrrolidone, acrylamide, methacrylic and maleic acids resulting in formation of water-soluble copolymers with a random distribution of the comonomer units in a macromolecule. *N,N*-diallyl-*N'*-acetylhydrazine is less active if compared with vinyl monomers. *N,N*-diallyl-*N'*-acetylhydrazine exhibits high activity at copolymerization with sulphur dioxide. The reaction results in alternating of copolymers of equimolar composition irrespective of reaction conditions. NMR and electron spectroscopy data proved *N,N*-diallyl-*N'*-acetylhydrazine to copolymerize (both double bonds are involved) with formation of pyrrolidine structures. Cyclolinear copolymers obtained are soluble owing to intramolecular cyclization of *N,N*-diallyl-*N'*-acetylhydrazine when formation of the polymer chain and to the absence of intermolecular crosslinks. Kinetic regularities of the process, structure and some properties of copolymers have been studied.

The interest to purposive synthesis of water-soluble polymers is growing steadily. It is conditioned by a wide set of the useful properties of mentioned polymers [1, 2]. N, S-containing water-soluble polymers, that proved to be promising flocculants [3], catalysts for a

* E-mail: vorobjeva@anrb.ru (Vorob'eva Antonina Ivanovna)

number of chemical syntheses [4], reagents for leather production [5], means for increasing the crop capacity [6], carriers of physiologically active substances etc, present particular interest.

However, inspite of such a wide application field, the assortment of used water-soluble polymers is limited. Therefore, a task of creation of the new types of water-soluble polymers based on available raw material, is a quite topical one. N-allyl derivatives of hydrazine, a product requiring utilization, are promising from this point of view.

Results of investigations concerning synthesis of water-soluble copolymers of N,N-diallyl-N'-acetylhydrazine with acrylamide, acrylic and maleic acids, N-vinylpyrrolidone and sulphur dioxide are shown in present paper.

EXPERIMENTAL

N,N-diallyl-N'-acetylhydrazine (DAAH) was obtained as a result of interaction of acetic acid hydrazide and allyl chloride by technique [7]. A fraction with T_b=121-125°C/5 mm Hg, n_D^{20}= 1,4770 was used in experiments.

N-vinylpyrrolidone (VP) was dried by potassium hydroxide and purified by vacuum distillation. A fraction with T_b= 97°C/13 mm Hg, n_D^{20}= 1,5117 was used.

Acrylic acid (AA) was dried by anhydrous calcium chloride and twice distilled under vacuum; a fraction with T_b=40°C/22 mm Hg, n_D^{20}= 1,4224 was used in the work.

Acrylamide (AAM) and maleic acid (MA) were recrystallized from acetone, dried under vacuum to a constant weight, T_m= 84 and 131°C, respectively.

Sulphur dioxide was dried by passing through concentrated sulphuric acid and just-sintered $CaCl_2$.

Characteristics of applied initiators (potassium persulphate (PP), benzoyl peroxide (BP), 2,2'-azobisisobutyronitrile (AIBN)) and solvents (DMSO, methanol, acetone, chloroform) conformed to the reference data after purification by conventional methods.

Copolymerization of DAAH was conducted in bulk and in solution at the presence of initiators: AIBN at conducting of the reaction in bulk and in organic solvent and PP – in aqueous medium. Kinetic investigations were carried out at initial conversions by gravimetric method at 60-90°C.

Copolymerization of DAAH with vinyl monomers was carried out in ampules under vacuum, and with SO_2 – in a glass reactor according to the following technique. Necessary quantity of SO_2 was introduced into liquid nitrogen -cooled reactor via recondensation stage, then necessary quantity of DAAH, initiator and solvent was added. Then reactor was shut and the reaction was conducted at the chosen temperature. In a certain interval (both in ampule and reactor) copolymer was precipitated and purified by three-fold reprecipitation by a precipitant from solution. Purified copolymers were dried under vacuum at 50°C until constant weight was attained. Copolymer composition was calculated from the data of element analysis.

UV-spectra were recorded on a "Shimadzu UV-VIS-NIR 3100" spectrometer. Complex formation was studied by deviations from additivity of absorbance differences of the monomer mixture solutions and the sums of absorbances of each comonomer at the same concentrations. The complex composition was determined by the method of isomolar series [8].

NMR ^{13}C spectra were registered on a "Bruker AM-300" spectrometer operating at 75,47 MHz using a broad-band proton suppression and in a JMOD mode. D_2O, $CDCl_3$ and DMSO-d_6 were used as solvents; 2,2-dimethyl-2-silapentan-5-sulphoasid (DSS) and tetramethylsilane (TMS) were used as internal standards.

RESULTS AND DISCUSSION

Allyl compounds, destinguished by a tendency to degradative chain transfer to the monomer, are known to exhibit low activity in reactions of radical polymerization. Therefore, the first stage of investigation was to find out if DAAH involvement into the processes of radical polymerization is possible. Investigations carried out showed, that DAAH does not homopolymerize according to free radical mechanism (at the presence of AIBN, PP, PB) at 60-90°C. DAAH activity somewhat increases in the case of reactions, conducted at the presence of protic and aprotic acids. However, in this case, particularly at polymerization at the presence of HCl, H_3PO_4, $ZnCl_2$ in the ratio DAAH:acid=1.1:1.2 ([AIBN]=3%, T=90°C, 30 hours), the polymer yield does not exceed 6 – 10%.

Being virtually inert at homopolymerization, DAAH appeared to copolymerize with vinyl monomers at the presence of radical initiators. The results of investigation on DAAH copolymerization with acrylic and maleic acids, N-vinylpyrrolidone and acrylamide, resulting in obtaining of water-soluble polymers, have been given in present paper. Copolymerization reactions proceed both in bulk and in solution with formation of copolymers, characterized by a random distribution of the comonomer units in a macromolecule (Table 1). DAAH is less active if compared with vinyl monomers (VM) - copolymers are enriched by VM units at all ratios of the monomers in the initial mixture. In particular, DAAH content in copolymers at initial equimolar monomer ratio is 15-20 mol % (Table 1). When conducting reaction at 80-90°C and initiator concentration 2.5-3.0 mas.% at equimolar monomer ratio the reaction rate of investigated systems is 2.7-4.1% per hour. The increase of DAAH fraction in the initial mixture allows to increase its content in the copolymer, but at the same time reaction rate decreases.

A DAAH-maleic acid system is an exclusion. Peculiarity of the system is a high tendency of the comonomer units toward alternation in the polymer chain (Table 1).

DAAH is significantly more active at copolymerization with sulphur dioxide, exhibiting high electron-acceptor activity. A study of copolymerization of DAAH with SO_2 showed the content of copolymers obtained to be of no dependence on the monomers ratio (Table 2), reaction conditions - temperature (Table 3), nature of initiator (Table 4), medium (Table 5), polymerization degree (Table 6) and to correspond to DAAH:SO_2 ratio 1:1.

Table 1. Copolymerization of DAAH with vinyl monomers (M_2) [AIBN] = 3.0 mass%., T = 90°C, 10 hours

M_2	Composition of the initial mixture, molar fraction		Medium	Yield %	Copolymer composition, molar fraction		Copolymer solvents
	M_1	M_2			M_1	M_2	
VP	0.50 0.55	0.50 0.45	In bulk	28.2 10.1	0.19 0.23	0.81 0.77	H_2O, methanol, acetone
AA	0.50 0.62	0.50 0.38	In bulk	27.2 16.3	0.15 0.22	0.85 0.78	H_2O, DMSO
MA	0.50 0.60	0.50 0.40	Methanol	18.9 7.8	0.49 0.50	0.51 0.50	H_2O, methanol
AAM	0.50 0.64	0.50 0.36	H_2O* Methanol	41.0 21.3	0.20 0.28	0.80 0.72	H_2O, DMSO

*Initiator - potassium persulphate

Table 2. Dependence of composition of DAAH copolymer with SO_2 at initial conversions on the monomer ratio. In bulk, [AIBN]=3.0 mass%, T=80°C

№	Composition of the initial mixture, mol %		Yield, %	Copolymer composition, mol %	
	DAAH	SO_2		DAAH	SO_2
1	20	80	8.8	49.8	50.2
2	50	50	10.2	50.1	49.9
3	80	20	8.9	50.0	49.0

Table 3. Dependence of composition of DAAH copolymer with SO_2 on the temperature. In bulk, DAAH:SO_2 = 1:1, [AIBN] = 3.0 mass%

№	Reaction temperature, ^{0}C	Yield, %	Polymer composition, mol %	
			ДААГ	SO_2
1	60	6.8	50.0	50.0
2	70	8.5	50.2	49.8
3	80	10.2	50.1	49.9
4	90	11.5	50.0	50.0

Table 4. Dependence of the DAAH with SO_2 copolymer composition on the initiator nature at initial conversions. DAAH: SO_2=1:1, T=80^{0}C

№	Initiator	Initiator concentration, mass%	Polymerization conditions	τ, min.	Yield, %	Copolymer composition, mol %	
						DAAH	SO_2
1	ДАК	2.0	In bulk	60	10.2	50.1	49.9
2	ПБ	3.0	In bulk	360	7.3	51.1	48.9
3	ПК	0.2	In water	20	12.2	49.7	50.3

Table 5. The influence of the medium on the composition of DAAH copolymers with SO_2 at initial conversions. $[M_1+M_2]$: solvent=50:50 mass%, [AIBN]=3.0 mass%, T=80^{0}C

№	Medium	Yield, %	Copolymer composition, mol %	
			DAAH	SO_2
1	In bulk	10.2	50.1	49.9
2	Benzene	9.9	50.3	49.7
3	DMSO	9.7	49.8	50.2
4	Acetone	8.4	50.0	50.0
5	Chloroform	13.0	51.4	49.6
6*	Water	12.2	49.7	50.3

*Potassium persulfate was used as initiator.

Table 6. The influence of conversion on the copolymer DAAH with SO_2 composition. DAAH: SO_2=1:1. $[M_1+M_2]$: solvent=50:50 mass %

Medium	Initiator, mass %	Temperature, ^{0}C	Yield, %	Copolymer composition, mol %	
				DAAH	SO_2
DMSO	AIBN; 2.0	90	9.8	50.2	49.8
	AIBN; 2.0	90	29.4	49.6	50.4
	AIBN; 2.0	90	47.7	49.3	50.7
Water	PP; 0.5	70	9.5	50.1	49.9
	PP; 0.5	70	30.4	50.0	50.0
	PP; 0.5	70	71.2	49.8	50.2

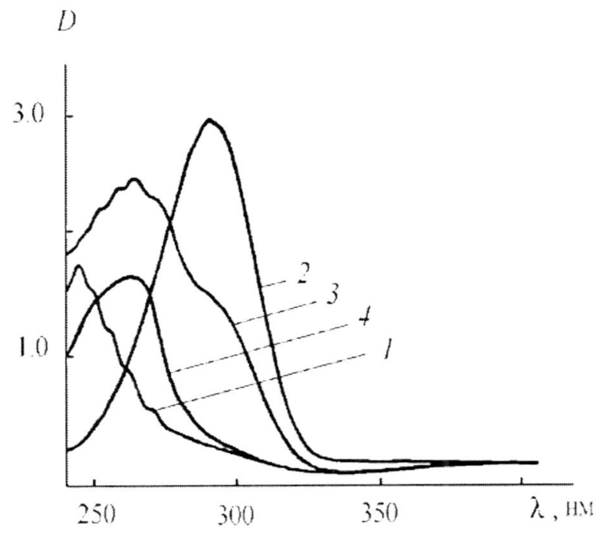

Figure1. Electron absorption spectra of solutions in chloroform: [DAAH] = 1.0 × 10^{-2} mol/l (1); [SO_2] = 1.0 × 10^{-3} mol/l (2); [DAAH] = 0.5 × 10^{-2} and [SO_2] = 0.5 × 10^{-3} mol/l mixture (3); spectrum (4) – difference of spectrum (3) and half-sum of specrta (1) и (2).

Composition constancy regardless of the monomer ratio in the reaction mixture allowed to suppose, that copolymerization of DAAH with SO_2 proceed via formation of the complexes. Investigations showed simple mixing of the comonomers at 20-80°C to result in momentary formation of a viscous amber-coloured adduct. A new band of a charge transfer with λ_{max}=263 nm (Figure 1) was detected in UV-spectra of DAAH and SO_2 mixture, recorded in chloroform. Formation of the complex of DAAH with SO_2, becoming apparent as a result of a deviation from additivity of absorbances of DAAH, SO_2 and their mixture (Figure 2), was also detected in DMSO and aqueous solutions. It is seen, that deviation maximum is observed at equimolar ratio of the monomers.

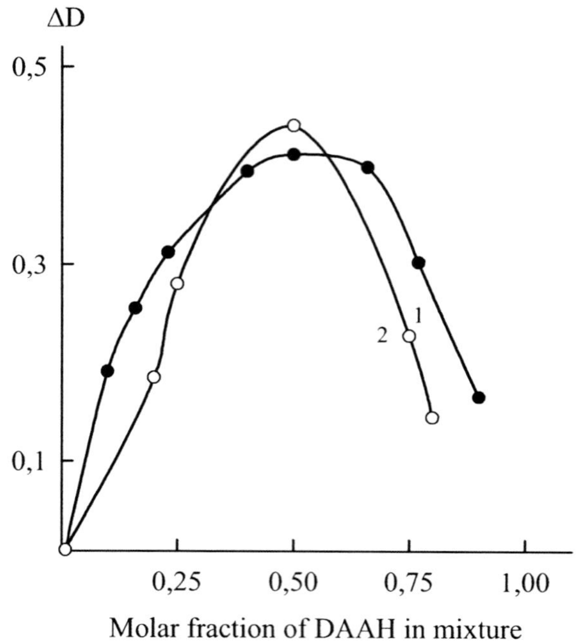

Figure 2. The absorbance difference of solutions of DAAH and SO_2 mixtures and the sum of absorbance of individual solutions of DAAH and SO_2 in water at λ = 277 nm (1) and DMSO at λ = 282 nm (2) vs. their ratio.

Comparison of DAAH and its adduct with SO_2 NMR ^{13}C spectra shows the end C_3 and C_3' atoms to shift to the weak field area substantially (~9 ppm), as well as the signals of atoms C_2 and C_2' to shift to the strong field area (~1 ppm). The shifts of the signals of the double bond carbon atoms denote distribution of π-electron density in DAAH molecule and formation of donor-acceptor complex [DAAH···SO_2]. Magnetic equivalence of the carbon atoms of allyl groups indicates the complex structure to be symmetric.

Reaction conditions significantly influence on the reaction rate and copolymer yield, having no influence on the copolymers composition.

A study of kinetic regularities of copolymerization in the aqueous media and DMSO solution showed an extremal dependence of reaction rate on the monomer ratio to be observed. The maximum rate was observed at equimolar ratio of monomers (Figure 3). Deviation of the comonomers ratio from equimolar both to higher and lower values of one of comonomers results in reaction rate reduction, the fact being explained probably by the change of the reactive complex concentration in the system.

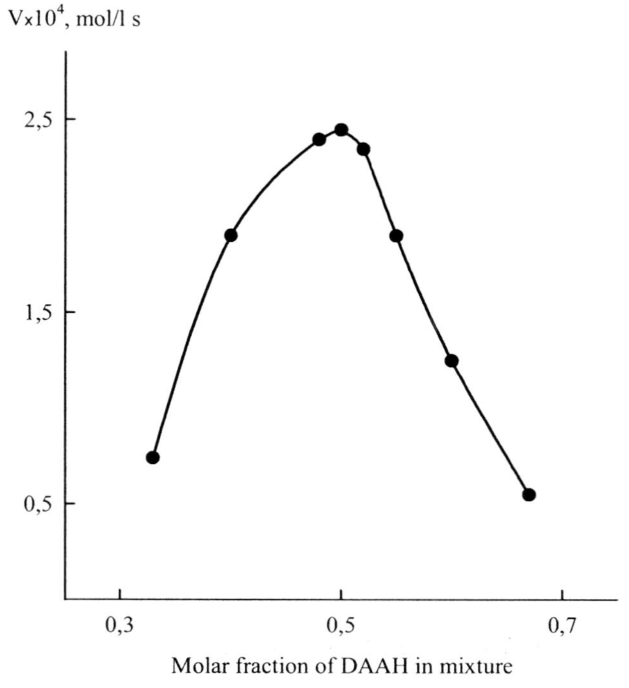

Figure 3. The rate of copolymerization of DAAH with SO_2 vs. the monomer ratio. DMSO; $[M_1 + M_2] = 5.3$ mol/l, $[AIBN] = 1.2 \times 10^{-1}$ mol/l; T = 90°C.

As a result of kinetic investigations at initial conversions, it was determined, that independently of medium (water/DMSO) a usual for radical polymerization reaction half order with respect to initiator is observed, indicating bimolecular mechanism of the growing chain failure, as well as deficiency of degradative chain transfer to the monomer, intrinsical to allyl monomers.

Reaction order with respect to the monomer appeared to be significantly dependent on the media. For example, in the case of reaction proceeding in the aqueous media, a unity order with respect to the monomer (with respect to the monomer sum at their equimolar ratio) typical of the radical polymerization reaction is observed, whereas it is equal to 0.71 in DMSO medium. Such a difference in the values of reaction order with respect to the monomer is probably connected with different stability of the formed $[DAAH \cdots SO_2]$ complex in aqueous and organic media, confirmed by spectral ($NMR^{13}C$) investigations of the complex (Table 7) - i.e. by different values of the signal shifts of carbon atoms of allyl groups of DAAH of obtained in aqueous and DMSO media complexes.

Table 7. NMR ^{13}C spectra of DAAH, DAAH complexes with SO_2 (in H_2O, DMSO***) and DAAH copolymers (D_2O, DSS, 25°C) (*-stereoisomers: c-cis, t-trans)**

	Formula	*	Chemical shift values and signals multiplets of the atoms (δ, ppm)										
			C_1,C_1'	C_2,C_2'	C_3,C_3'	C_4	C_5	C_6	C_7	C_8	C_9	C_{10}	C_{11}
1	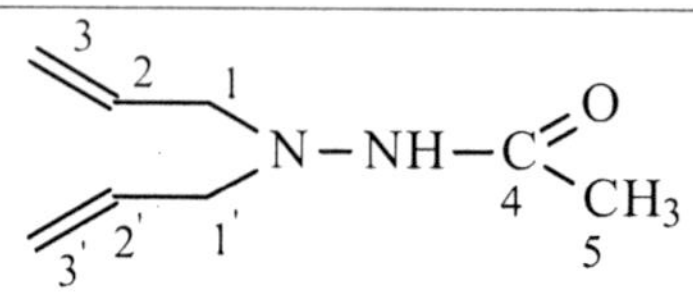		59.11 t	132.12 d	117.14 t	172.15 s	18.69 q						
2	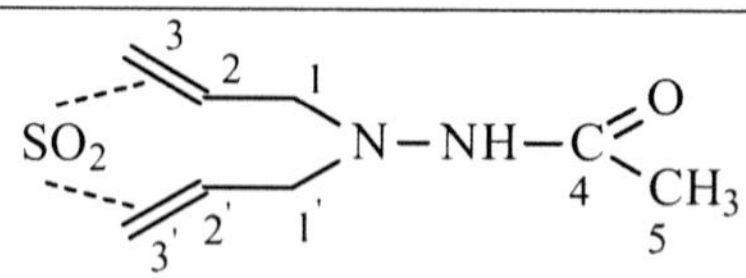 **		62.79 t	131.16 d	126.51 t	174.24 s	22.24 q						
3	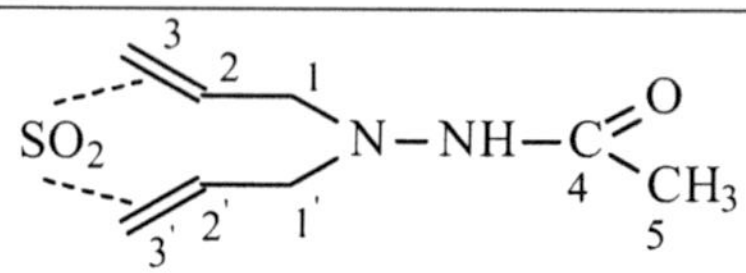 ***		60.49 t	134.58 d	118.72 t	173.22 s	20.40 q						
4	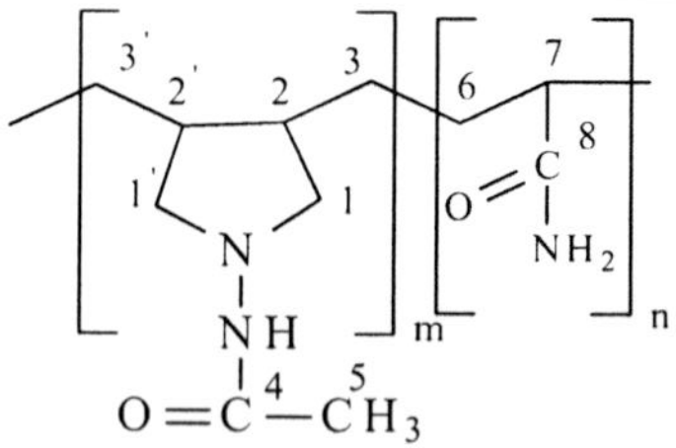	c t	62.20 62.35 t	40,19 40,88 d	28,09 28,41 t	175.04 174.75 s	21.69 q	36.19 37.14 36.82 t	44.40 44.18 43.82 d	182.09 183.05 s			

Continuation of the Table 7.

	Formula	*	Chemical shift values and signals multiplets of the atoms (δ, ppm)										
			$C_1,C_{1'}$	$C_2,C_{2'}$	$C_3,C_{3'}$	C_4	C_5	C_6	C_7	C_8	C_9	C_{10}	C_{11}
5	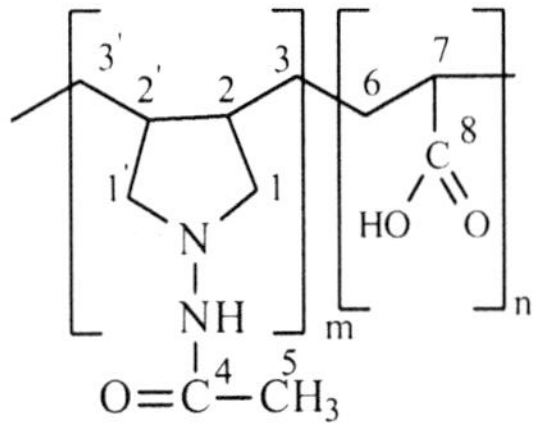	c t	58.76 59.72 t	34.80 36.05 d	44.38 T	176.74 s	19.47 q	43.85 t	51.60 d	177.15 176.17 s			
6	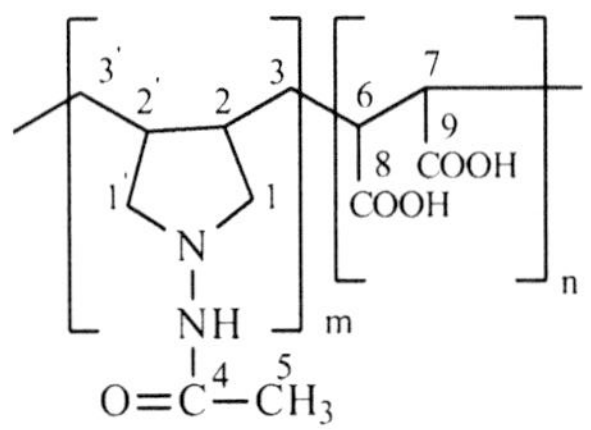	c t	58.89 60.63 t	46.71 48.05 d	33.60 35.99 T	172.03 s	21.03 q	51.54 d	51.54 d	173.55 s	173.55 s		
7	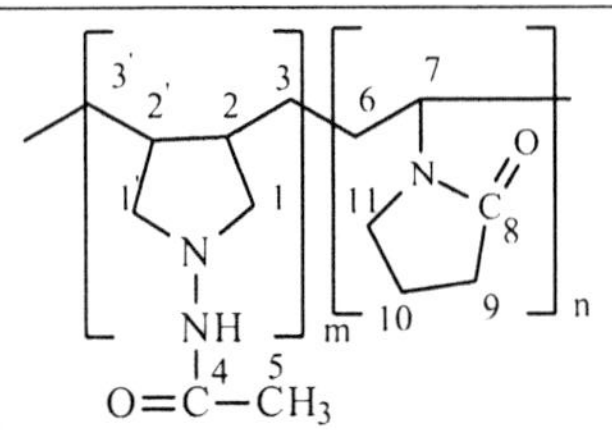	c t	62.25 64.85 t	35.08 40.63 d	29.14 32.14 t	173.54 s	21.45 q	31.85 t	46.25 48.06 d	179.78 s	33.67 t	19.95 t	44.76 t
8	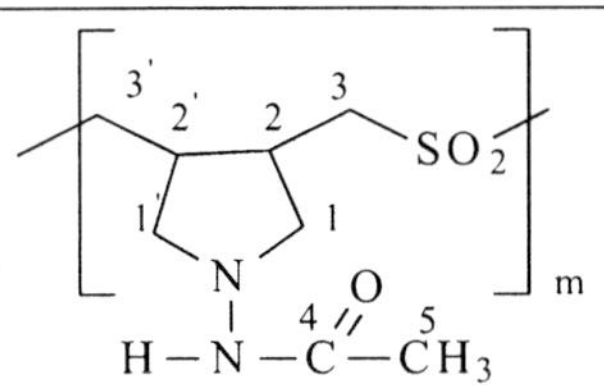	c t	61.90 61.33 t	35,41 37,82 d	54,01 57,02 t	177.13 s	22.42 q						

The reaction rate in the studied range (60-90°C) increases with temperature independently of the media. DAAH copolymerization with SO_2 proceeds significantly easier in aqueous media - at lower temperatures and lesser concentrations of initiator, than in DMSO. The values of the total activation energy of DAAH copolymerization with SO_2 in the water solution and DMSO, calculated from Arrenius equation, are 58,6 and 87,9 kJ/mol, respectively. It is known from reference data [9] that the values of activation energy of alternate copolymerization turn out to be lesser, than in the case of the most reactions of free-radical polymerization, effective activation energy of the latter being in the range 83,7-96,3 kJ/mol.

It is seen from Table 4, that the higher reaction rates are observed when using of AIBN and PP as initiators. Copolymerization virtually does not proceed when using bensoyl peroxide as an initiator.

Polymer yield at copolymerization in bulk does not exceed 50% even at exhaustively long (to 10 hours) duration of the reaction, this fact being peculiar for the system considered. The copolymer yield amounts to 70-80% at conducting of the reaction in solution (H_2O, DMSO, chloroform).

There are two dominant sets of signals in NMR ^{13}C spectra of copolymers of DAAH with VM (Table 7), the values of chemical shifts (c.s.) of these sets, including the signals of configurational multiplets of dyads, virtually coinsiding with the values of the homopolymers (PAAM, PVP, PAA) chemical shifts. Besides, in carbon spectra of the copolymers there are low intensity triplet signals, corresponding to C_3 and C_6 atoms of the joint comonomer units. This fact indicates, that copolymers contain blocks of VM units, divided by DAAH units.

Signals of the carbon atoms (C_6-C_9), related to the MA units in copolymers, are appreciably widened, that is connected with the presence of intramolecular hydrogen bonds between COOH-groups of MA and neighbour -C=O and NH- groups of DAAH.

In the spectra of copolymers of DAAH with SO_2 there are two couples of the weak field triplet signals of stereoisomeric C_3, C_3' (54.01; 57.02 ppm) and C_1, C_1'(61.90, 61.33 ppm), a couple of doublet signals of C_2, C_2' atoms (35.41, 37.82 ppm), c.s. values close to those given in [10] for copolymers of N,N-dimethyl-N,N-diallylammonium chloride with SO_2, and two signals of acetyl group.

Analysis of presented copolymer spectra shows DAAH to copolymerize with VM and SO_2, both double bonds participating, with formation of cis-, trans-stereoisomeric pyrrolidine structures in cyclolinear polymer chain in the proportion ~4:1.

Copolymers of DAAH with SO_2 in the form of white powder are soluble in water, DMSO and DMFA and insoluble in the other widely used solvents (acetone, MEK, methanole, benzene, dioxane, TGF, ethylacetate, chlorinated hydrocarbons and oth.). Solubility of the DAAH copolymers with VM depends on the nature of the second comonomer; copolymer solvents are listed in Table 1.

Preliminary tests showed copolymers of DAAH to have a high flocculating activity at precipitation of $Cu(OH)_2$.

Thus, DAAH virtually does not homopolymerize by free-radical mechanism, but does copolymerize with VM and MA with formation of watersoluble copolymers with a random distribution of the comonomer units in the molecule. DAAH is less active compared with VM, copolymers are enriched by VM units in the whole interval of the monomer ratios. The rate of copolymerization reaction with VM decreases with the increase of the DAAH fraction in the initial mixture. DAAH exhibits a high activity at copolymerization with sulphur dioxide. Copolymerization proceeds via complexation resulting in obtaining of alternating copolymers of equimolar composition undependently of the monomer ratio in the initial mixture and reaction conditions. DAAH is involved in copolymerization reactions with participation of the both double bonds and result in formation of pyrrolidine structures in the cyclolinear polymer chain.

Financial support by the Russian Fund for Basic Reseach (grant № 05-03-32097) is gratefully acknowleaged.

References

[1] Kirsh Yu.E. Poly-N-vinylpyrrolidone and other poly-N-vinylamides: *Synthesis and physico-chemical properties.* – M.: Nauka, 1998. – 252 p.

[2] Boyarkina N.M., Kryuchkov V.V., Parkhamovich E.S., Amburg L.A., Topchiev D.A., Kabanov V.A. // *Plast.massy.* – 1987. – № 8. – p. 17.

[3] Andreson R.K., Leplyanin G.V., Vorob'eva A.I., Sakhautdinov V.G., Speranskiy V.V., Propadushchaya L.A., Andreson B.A., Topchiev D.A. USSR Inventor's Certificate № 1345696 // *Bull. Izobret.* 1987. – № 38. – p.271.

[4] Leplyanin G.V., Tolstikov G.A., Vorob'eva A.I., Shurupov E.V., Abdrashitov Ya.M., Bikbaeva G.G., Sysoeva L.B., Sataeva F.A., Kozlov V.G. USSR Inventor's Certificate № 1530631 // *Bull. Izobret.* 1989. – № 47. – p. 122.

[5] RF Patent № 2145978 // *Bull. Izobret.* 2000. – № 6. – p. 235.

[6] Gilyazetdinov Sh.Ya., Balakhontsev E.N., Leplyanin G.V., Vorob'eva A.I., Iskhakov F.F., Radtseva O.V. USSR Inventor's Certificate № 1744797 // *Bull. Izobret.* 1992. – № 24. – p.205.

[7] Gusev V.Yu., Vorob'eva A.I., Radushev A.V., Kolesov S.V., Muslukhov R.R., Tolstikov A.G. // *Dokl. Akad. Nauk, Ser. Khim.* – 2003. – № 12. – p. 2603.

[8] Gur'yanov E.N., Gol'dshtein I.P., Romm I.P. Donor-acceptor bond. – M.: *Khimiya,* 1973. – 397 p.

[9] Kabanov V.A., Zubov V.P., Semchikov Yu.D. Complex- radical polymerization. – M.: *Khimiya,* 1987.

INDEX

A

B

C

D

E

F

G

H

I

J

K

L

M

N

O

P

Q

R

S

T

U

V

W

Y

Z